TREATISE ON BASIC PHILOSOPHY

Volume 1

SEMANTICS I: SENSE AND REFERENCE

TREATISE ON BASIC PHILOSOPHY

1

SEMANTICS I *Sense and Reference*

2

SEMANTICS II *Interpretation and Truth*

3

ONTOLOGY I *The Furniture of the World*

4

ONTOLOGY II *A World of Systems*

5

EPISTEMOLOGY I *The Strategy of Knowing*

6

EPISTEMOLOGY II *Philosophy of Science*

7

ETHICS *The Good and the Right*

MARIO BUNGE

Treatise on Basic Philosophy

VOLUME 1

Semantics I:

SENSE AND REFERENCE

D. REIDEL PUBLISHING COMPANY
DORDRECHT-HOLLAND / BOSTON-U.S.A.

Library of Congress Catalog Card Number 74–83872

ISBN 90 277 0534 8

Published by D. Reidel Publishing Company,
P.O. Box 17, Dordrecht, Holland

Sold and distributed in the U.S.A., Canada, and Mexico
by D. Reidel Publishing Company, Inc.
306 Dartmouth Street, Boston,
Mass. 02116, U.S.A.

Printed in The Netherlands by D. Reidel, Dordrecht

GENERAL PREFACE TO THE *TREATISE*

This volume is part of a comprehensive *Treatise on Basic Philosophy*. The treatise encompasses what the author takes to be the nucleus of contemporary philosophy, namely semantics (theories of meaning and truth), epistemology (theories of knowledge), metaphysics (general theories of the world), and ethics (theories of value and of right action).

Social philosophy, political philosophy, legal philosophy, the philosophy of education, aesthetics, the philosophy of religion and other branches of philosophy have been excluded from the above *quadrivium* either because they have been absorbed by the sciences of man or because they may be regarded as applications of both fundamental philosophy and logic. Nor has logic been included in the *Treatise* although it is as much a part of philosophy as it is of mathematics. The reason for this exclusion is that logic has become a subject so technical that only mathematicians can hope to make original contributions to it. We have just borrowed whatever logic we use.

The philosophy expounded in the *Treatise* is systematic and, to some extent, also exact and scientific. That is, the philosophical theories formulated in these volumes are (*a*) formulated in certain exact (mathematical) languages and (*b*) hoped to be consistent with contemporary science.

Now a word of apology for attempting to build a system of basic philosophy. As we are supposed to live in the age of analysis, it may well be wondered whether there is any room left, except in the cemeteries of ideas, for philosophical syntheses. The author's opinion is that analysis, though necessary, is insufficient – except of course for destruction. The ultimate goal of theoretical research, be it in philosophy, science, or mathematics, is the construction of systems, i.e. theories. Moreover these theories should be articulated into systems rather than being disjoint, let alone mutually at odds.

Once we have got a system we may proceed to taking it apart. First the tree, then the sawdust. And having attained the sawdust stage we should move on to the next, namely the building of further systems. And this for three reasons: because the world itself is systemic, because no idea can become fully clear unless it is embedded in some system or other, and because sawdust philosophy is rather boring.

The author dedicates this work to his philosophy teacher

Kanenas T. Pota

in gratitude for his advice: "Do your own thing. Your reward will be doing it, your punishment having done it".

CONTENTS OF *SEMANTICS I*

PREFACE TO *SEMANTICS I*

This is a study of the concepts of reference, representation, sense, truth, and their kin. These semantic concepts are prominent in the following sample statements: ⌜The field tensor *refers* to the field⌝, ⌜A field theory *represents* the field it refers to⌝, ⌜The *sense* of the field tensor is sketched by the field equations⌝, and ⌜Experiment indicates that the field theory is *approximately true*⌝.

Ours is, then, a work in philosophical semantics and moreover one centered on the semantics of factual (natural or social) science rather than on the semantics of either pure mathematics or of the natural languages. The semantics of science is, in a nutshell, the study of the symbol-construct-fact triangle whenever the construct of interest belongs to science. Thus conceived our discipline is closer to epistemology than to mathematics, linguistics, or the philosophy of language.

The central aim of this work is to constitute a semantics of science – not any theory but one capable of bringing some clarity to certain burning issues in contemporary science, that can be settled neither by computation nor by measurement. To illustrate: What are the genuine referents of quantum mechanics or of the theory of evolution?, and Which is the best way to endow a mathematical formalism with a precise factual sense and a definite factual reference – quite apart from questions of truth?

A consequence of the restriction of our field of inquiry is that entire topics, such as the theory of quotation marks, the semantics of proper names, the paradoxes of self-reference, the norms of linguistic felicity, and even modal logic have been discarded as irrelevant to our concern. Likewise most model theoretic concepts, notably those of satisfaction, formal truth, and consequence, have been treated cursorily for not being directly relevant to factual science and for being in good hands anyway. We have focused our attention upon the semantic notions that are usually neglected or ill treated, mainly those of factual meaning and factual truth, and have tried to keep close to live science.

The treatment of the various subjects is systematic or nearly so: every basic concept has been the object of a theory, and the various theories have been articulated into a single framework. Some use has been made of certain elementary mathematical ideas, such as those of set, function, lattice, Boolean algebra, ideal, filter, topological space, and metric space. However, these tools are handled in a rather informal way and have been made to serve philosophical research instead of replacing it. (Beware of hollow exactness, for it is the same as exact emptiness.) Moreover the technical slices of the book have been sandwiched between examples and spiced with comments. This layout should make for leisurely reading.

The reader will undoubtedly apply his readermanship to skim and skip as he sees fit. However, unless he wishes to skid he will be well advised to keep in mind the general plan of the book as exhibited by the Table of Contents. In particular he should not become impatient if truth and extension show up late and if analyticity and definite description are found in the periphery. Reasons will be given for such departures from tradition.

This work has been conceived both for independent study and as a textbook for courses and seminars in semantics. It should also be helpful as collateral reading in courses on the foundations, methodology and philosophy of science.

This study is an outcome of seminars taught at the Universidad de Buenos Aires (1958), University of Pennsylvania (1960–61), Universidad Nacional de México (1968), McGill University (1968–69 and 1970–71), and ETH Zürich (1973). The program of the investigation and a preview of some of its results were given at the first conference of the Society for Exact Philosophy (see Bunge, 1972a) and at the XVth World Congress of Philosophy (see Bunge, 1973d).

ACKNOWLEDGEMENTS

It is a pleasure to thank all those who made useful comments and criticisms, whether constructive or destructive, in the classroom or in writing. I thank, in particular, my former students Professors Roger Angel and Charles Castonguay, as well as Messrs Glenn Kessler and Sonmez Soran, and my former research associates Professors Peter Kirschenmann, Hiroshi Kurosaki, Carlos Alberto Lungarzo, Franz Oppacher, and Raimo Tuomela, and my former research assistants Drs David Probst and David Salt. I have also benefited from remarks by Professors Harry Beatty, John Corcoran, Walter Felscher, Joachim Lambek, Scott A. Kleiner, Stelios Negrepontis, Juan A. Nuño, Roberto Torretti, Ilmar Tammelo, and Paul Weingartner. But, since my critics saw only fragments of early drafts, they should not be accused of complicity.

I am also happy to record my deep gratitude to the Canada Council for the Killam grant it awarded this research project and to the John Simon Guggenheim Memorial Foundation for a fellowship during the tenure of which this work was given its final shape. Finally I am grateful to the Aarhus Universitet and the ETH Zürich for their generous hospitality during my sabbatical year 1972–73.

MARIO BUNGE

Foundations and Philosophy of Science Unit,
McGill University

SPECIAL SYMBOLS

C Set of constructs (concepts, propositions, or theories)
$\mathbb{C}$ Context
$\mathcal{C}$ Content (extralogical import)
$\mathcal{C}n$ Consequence
$\mathcal{D}$ Designation
Δ Denotation
$\hat{=}$ Representation
$\mathcal{E}$ Extension
$\mathcal{I}$ Intension
$\mathcal{I}mp$ Import (downward sense)
L Logic
$\mathcal{L}$ Language
$\mathcal{M}$ Meaning
Ω Universe of objects (of any kind)
$\mathbb{P}$ Family of predicates
$\mathcal{P}ur$ Purport (upward sense)
$\mathcal{R}$ Reference
S Set of Statements (propositions)
$\mathcal{S}$ Sense
$\mathcal{S}ig$ Signification
T Theory (hypothetico-deductive system)
$\mathcal{V}$ Truth value function

INTRODUCTION

In this Introduction we shall sketch a profile of our field of inquiry. This is necessary because semantics is too often mistaken for lexicography and therefore dismissed as trivial, while at other times it is disparaged for being concerned with reputedly shady characters such as meaning and allegedly defunct ones like truth. Moreover our special concern, the semantics of science, is a newcomer – at least as a systematic body – and therefore in need of an introduction.

1. Goal

Semantics is the field of inquiry centrally concerned with meaning and truth. It can be empirical or nonempirical. When brought to bear on concrete objects, such as a community of speakers, semantics seeks to answer problems concerning certain linguistic facts – such as disclosing the interpretation code inherent in the language or explaning the speakers' ability or inability to utter and understand new sentences of the language. This kind of semantics will then be both theoretical and experimental: it will be a branch of what used to be called 'behavioral science'. Taking a cue from Chomsky and Miller (1963) we may say that, rather than being a closely knit and autonomous discipline, this kind of semantics is the union of two fields: a chapter of linguistics and one of psychology:

Empirical Semantics
- *Linguistic semantics:* the semantics of natural languages
- *Psycholinguistic semantics:* the psychology of speech acts and contents, or the study of language users

If, on the other hand, semantics concerns itself with conceptual objects, such as mathematical structures or scientific hypotheses, then it remains nonempirical in the sense that it makes no direct use of observation and measurement to test its conjectures and models – nor does it need to, because this kind of semantics does not describe and predict facts. In other words, nonempirical semantics is concerned not only with linguistic items but also, and primarily, with the constructs which some such items stand for as well as with their eventual relation to the real world. (More on constructs in Ch. 1, Sec. 1.2.) This branch of semantics is then closer to the theory of knowledge than the theory of language. (We shall take a look at this point in Ch. 10, Sec. 3.) Moreover, under penalty of being useless, nonempirical semantics should account for our experience with conceptual objects – thus being vicariously empirical. In particular, it should be concerned with our experience in interpreting conceptual symbols, elucidating the sense of constructs, uncovering their referents, and estimating truth values. Furthermore, the way it performs this task should constitute the supreme test of this branch of semantics: there is no justification for a semantic theory that fits neither mathematics nor science nor ordinary knowledge. Conceived in this way, nonempirical semantics may be split up as follows:

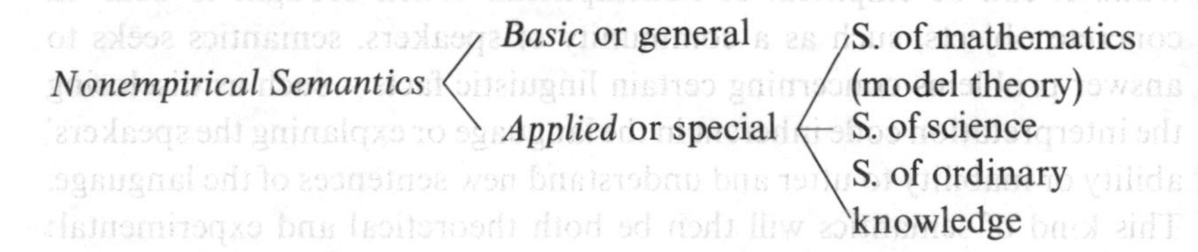

This book is concerned with but a fragment of nonempirical semantics and not at all with empirical semantics. The philosopher as such is not competent to tackle problems in empirical semantics: the best he can do is to study it with the hope of ferreting out the methodology and philosophy of scientific linguistics and psycholinguistics. Of course he may occasionally ask penetrating questions and, with less frequency, propose valuable insights in this field or any other. But, *qua* philosopher, he is not competent to expand such insights into theories proper (i.e. theories in mathematical linguistics), nor is he equipped to design experiments to put such theories to the test. The philosopher, as such, is as much of an amateur in linguistics and psychology as in physics and biology. But he can be a professional philosophical semanticist.

We shall deal exclusively with nonempirical semantics, i.e. with the semantic problems that cannot be investigated by empirical means because they are not concerned with any factual items but, at most, with certain features of our knowledge of such objects. In particular we shall not study the bewildering variety of concepts designated by the ambiguous term 'meaning' (see Schaff, 1962; Cohen, 1966; Hill, 1971). We shall instead restrict ourselves to the semantic concept of meaning, i.e. the sense and reference of predicates, propositions, and theories. In particular we shall not investigate the process by which an organism assigns a significance to a sign: we regard the pragmatic concept of meaning as a business of psycholinguistics (see e.g. Osgood *et al.*, 1957; Luria, 1969), anthropologists, and historians. Likewise we are interested in the semantic concept(s) of truth rather than in the psychological concepts of personal truth, strength of belief, credibility, and so on. Our choice does not imply a rejection of either empirical semantics or pragmatics: it only involves a deliberate restriction of the field of our inquiry – *ergo* a choice of method.

A further restriction on the scope of our investigation will be this. We shall pay special attention to the *semantics of factual science*, one of the three branches of applied or special semantics we recognized a while ago. That is, our inquiry shall focus on the notions of factual reference, factual sense, and factual truth that are relevant to scientific knowledge. The ultimate goal of our research is to accomplish for these notions what model theory has done for the concepts of satisfaction, formal truth, consequence, and extension – without however hoping to attain the neatness characterizing model theory. (For the semantics of logic see van Fraassen (1971), for the semantics of mathematics consult Robinson (1963) or Bell and Slomson (1969).)

In the pursuit of our goal we shall attempt to be both systematic and relevant to real science. To put it in nasty terms: We shall do our best to avoid the two chief blemishes of most of the available views and theories in basic semantics. These shortcomings are the lack of comprehensive systems in which *all* of the semantic concepts hang together and thus illumine one another, and the *irrelevance* to the semantic problems that pop up in live science. We shall try to give a fairly exact treatment but, if we have to choose between a fruitful insight on the one hand, and a rigorous but useless formalism on the other hand, we shall prefer the former. For, as cuckoos and physicists well know, given a fertilized egg there exists a bird willing to hatch it.

Because our semantic system is frankly unorthodox, its measure of success should not be gauged by its agreement with any of the extant views. It should be estimated, rather, by its ability to (*a*) clarify and codify hitherto obscure or stray notions, (*b*) perform suitable semantic analyses of fragments of current factual science, and (*c*) help in the systematic (axiomatic) reconstruction of existing theories in factual science.

2. Method

Because our primary concern will be with the semantics of scientific theories, our investigation will rank largely as metatheoretical. Now, every metatheory is couched in a metalanguage allowing us to express certain statements about items occurring in the (object) language used to express the propositions of the object theory. Throughout our quest we shall use as strong a metalanguage as required, for our aim is rather to get a few things done than to economize. We shall employ whatever tool may seem promising to attain precision, unity and clarity: in particular we shall use elementary set theory, a few algebraic theories, and a sprinkling of topology. (For a superb précis of the mathematics needed by the exact philosopher see Hartnett (1963, 1970).) In this sense our semantics will be in line with the semantics of Tarski and Carnap. It differs from theirs (*a*) in its *object*, which is factual science instead of mathematics, (*b*) in its *philosophical underpinnings*, which are realist rather than either nominalist, empiricist, of Fregean, and (*c*) in its *degree of formalization*, which is lower. While the first two differences are clear, the third deserves a short explanation.

Our exposé, though quite formal and systematic, will not be formalized in the metamathematical sense of the word. In particular we shall not specify our metalanguage in advance. We shall not do it partly because any such specification places a restriction, often unsuspected, upon the concepts and propositions that can be expressed in that language. And we do not care for such limitations, particularly when it is not clear which they are. For example, we want to be able to speak of infinite conjunctions and disjunctions, such as law statements, and even of nondenumerable sets of propositions, such as the formulas representing the successive positions of a particle in the course of time. To write off such constructs just because they overflow finitary logic would be unwise, and to try and find out in each case which logical theory justifies them would be going beyond the scope of this work. We take it for granted that, given a useful mathematical or scientific concept, there exists a branch of logic on which it can roost.

As stated above, we assume that logic and mathematics are sufficient to uncover the form and structure of every construct. We assume also that they are necessary to build theories aiming at elucidating and systematizing the semantic concepts of factual reference, factual sense, and factual truth. But, of course, we do not claim that logic, mathematics and the semantical theories contrived with their help will suffice to reveal the syntax and the semantics of every particular scientific construct – any more than geometry suffices to triangulate the universe. However, a combination of logic, mathematics, semantics and substantive knowledge can do the trick of uncovering the formalism and the semantics of a scientific theory. At any rate nothing else has done it, so it is worth trying.

The preceding assumptions concerning the role of logic and mathematics in constructing philosophical theories, such as semantics, characterize what may be called *exact philosophy* (see Bunge, ed., 1973a). They are likely to be challenged by philosophers of a traditional cast of mind. But they may also be doubted by progressives. A case in point is the notion of sense (or intension or connotation or content) – a favorite with conservatives as well as a *bête noire* of progressives for being allegedly impregnable to mathematization. We shall meet the challenge by proposing a mathematical theory of sense – or rather three such theories, one for each component of the sense of a construct (Chs. 4 and 5). Should these particular theories fail, others could take their place: approaches and programs die only if nobody works on them. After all, simpler ideas, like that of many, or class, were regarded as typically nonmathematical, hence unclear, until about one century ago. In any event, a bill of difficulties and failures does not constitute a philosophy.

Ordinary language philosophers and hermeneutic philosophers will complain that our method is misplaced, because logic and mathematics are incapable of discerning the structural subtleties of ordinary language (which one pray?). And if smart they will score a point of two, either because the locution in question has not yet been tamed or because the mathematicians have failed to pay proper attention to it. Given chance and motivation, mathematicians will tackle any problem of form: after all this is their province.

What holds for the analysis of form is also true of the analysis of meaning. Thus at first sight it would seem that 'or' and 'and' are not commutative in ordinary language: that their place makes a difference to meaning. If this were true then logic would be guilty of a gross oversimplification that would render it incompetent to analyze ordinary language. Example:

⌜You should study as well as play.⌝ (1)

does not mean the same as

⌜You should play as well as study.⌝ (2)

These statements are obviously different. Yet their ostensive structure, namely $s \& p$, and $p \& s$, is the same. Now, 'ostensive' signifies "superficial". There are hidden assumptions (presuppositions) that must be unearthed. In the case of (1) the presupposition is that the interlocutor, who is supposed to study, is neglecting his studies, whereas in the case of (2) the presupposition is that he is studying and neglecting his playing. More suitable formalizations would therefore be

$$(1) = \neg s \,\&\, p \,\&\, Obl\,(s \,\&\, p)$$
$$(2) = s \,\&\, \neg p \,\&\, Obl\,(s \,\&\, p) \neq (1),$$

where '*Obl*' stands for the deontic operator "ought", taken care of by deontic logic.

Conclusion: As with structure analysis so with meaning analysis: no genuine shade of meaning need remain uncovered if our analysis is pushed far enough. The resulting depth of our analysis will depend on the power of the analytic tools we have employed. What holds for logical form and for meaning holds for philosophical problems in general: if genuine they can be and ought to be tackled in an exact manner. Even though some exact philosophers may lack the *esprit de finesse* characterizing some inexact philosophers, the problems the latter discover can only be solved with a pinch of *esprit de géométrie.*

A final remark on method. We shall be concerned throughout with theories and their components, i.e. statements, or the designata of declarative sentences. Hence we need not investigate nonalethic constructs such as problems (expressed by questions or by commands) and norms (expressed by imperatives). In principle these and other non-declarative phrases may be treated by either of two methods: the direct and the indirect. The *direct* procedure consists in taking the bull by the horns and building theories (e.g. systems of deontic logic and of erotetic logic) legalizing, codifying and so cleansing some of our naive ideas on the subject. The *indirect* method consists in transforming the problem by translating the given nondeclarative sentence into a declarative one – i.e. in shedding the pragmatic trimmings of the original phrase. For example, 'Run!' may be transformed into 'The subject is ordered to run', and 'Where is x?' into 'The question is to locate x'. The problem whether every nondeclarative sentence has a "declarative prototype" (Marhenke, 1950) or a "propositional content" (Searle, 1969) seems open but does not affect our enterprise. We need not commit ourselves on this matter and we need not espouse either method for, as pointed out a while ago, scientific theories – our chief analysanda – contain just statements: only the research process leading to and from theories involves questions, norms, promises, threats, etc. However, our results concerning reference, sense, and truth, can be applied to nonalethic constructs provided the latter can be translated or distilled into their alethic counterparts. In this way we can speak of the meaning of problems and norms, or of the significance of questions and commands. In other words, we shall adopt the following principle: *If a construct has an alethic counterpart then the semantics of the former (though not its pragmatics) equals the semantics of the latter.*

Having delineated the goal and method of our endeavor, we proceed to take off.

CHAPTER 1

DESIGNATION

The goal of this chapter is to characterize the most basic of all semantic concepts: that of designation. This concept occurs in statements such as ⌜Sign x designates concept (or proposition) y⌝. Since a significant sign is a member of some language or other, we must start by defining the notion of a language and, in particular, that of a conceptual language, i.e. one capable of expressing propositions. But this will necessitate a clarification of the very nature and status of concepts and propositions. Which will in turn lead us to discussing some of the philosophical underpinnings and ramifications of our enterprise – for philosophical semantics, though a distinct discipline, is not an isolated one.

1. Symbol and Idea

1.1. *Language*

An artificial sign, whether written, uttered, or in any other guise, is a physical object – a thing or a process that a thing undergoes. But of course it is a very special object, namely one that

(i) *represents* some other object (physical or conceptual) or is part of some much proxy,

(ii) belongs to some *sign system* (= language), within which it can concatenate with other signs to produce further signs, and such that the whole system is used for

(iii) the *communication* or transmission of information concerning states of affairs, ideas, etc.

For something to be called a sign it need not be spoken or written: the signs used by bees and apes satisfy all the preceding conditions. Consequently a language need not be symbolic (i.e. involving conventions), let alone conceptual. Any *system of coded signals used for communication purposes* qualifies as a language. Moreover, the coding and decoding need not be accompanied by understanding: they can be automatic – as is often the case within humans. In any event we have the following possible partition of languages.

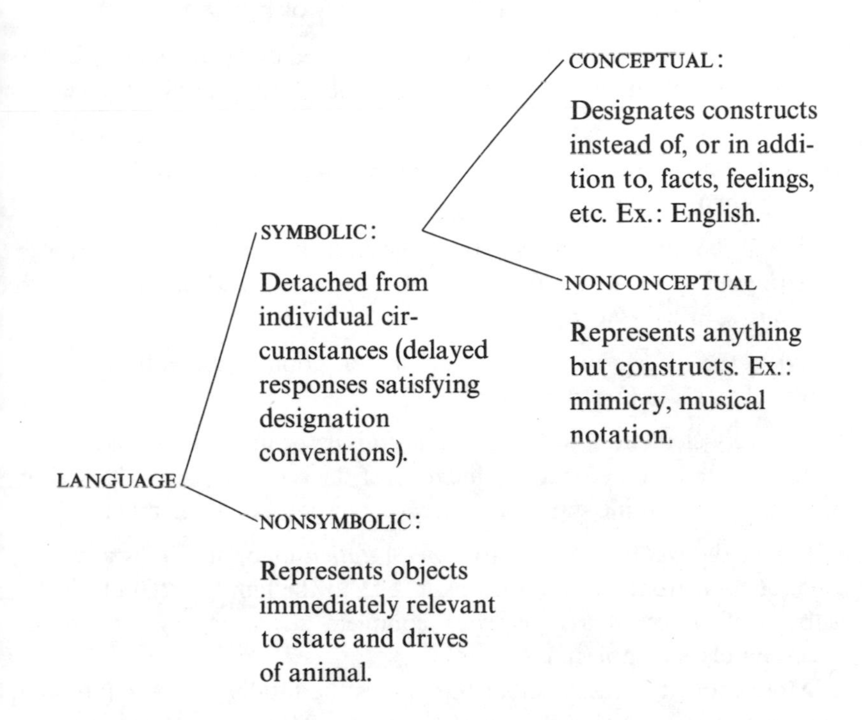

We are interested only in symbolic languages: the study of our various systems of growls, snarls, shrieks, and gestures belongs to psychology. (Moreover we are not interested in individual symbols but rather in equivalence classes of such: linguistics is concerned with the tee shape rather than with any concrete instance of the sound or letter tee.) Essentially, a symbolic language is a set of basic symbols (the alphabet) that can concatenate forming strings. A formation device picks the subset of expressions or well formed formulas, which in turn designate certain objects. And a transformation device converts certain expressions into others. More precisely, we shall adopt the following characterization of the overall structure of a finitary symbolic language (whether or not conceptual):

DEFINITION 1.1 A septuple $\mathscr{L}_K = \langle \Sigma, \Box, \circ, \phi, \tau, \Omega, \Delta \rangle$, where K, Σ and Ω are sets, $\Box$ is a distinguished element of Σ, $\circ$ is an operation on Σ, ϕ and τ are families of mappings, and Δ is a function, is called a (finitary) *symbolic language* for communicating systems of kind K iff

(i) Σ, the *alphabet* of $\mathscr{L}_K$, is denumerable and every element of Σ is a symbol that any member of the class K of things can produce or receive;

(ii) the structure $\langle \Sigma^*, \Box, \circ \rangle$, where Σ^* is the set of finite concatenations (*strings*) of elements of Σ, is the free monoid generated by Σ, with identity (neutral) element $\Box$ (the *blank*);

(iii) ϕ, the *formation* device, is a collection of maps from the n-tuples of strings in $(\Sigma^*)^n$ to a subset Σ^{**} of Σ^* (the *expressions*, or *complete utterances*, or *wff*'s of $\mathscr{L}_K$);

(iv) τ, the *transformation* device, is a collection of maps from the n-tuples of wff's in $(\Sigma^{**})^n$ into Σ^{**};

(v) every element of the set Ω of *objects* is associated with (or it evokes) a definite state in an arbitrary member of the set K of users of $\mathscr{L}_K$ (but not conversely: some states of all users correspond to no members of Ω);

(vi) Δ, the *coding* function (or *interpretation* map) of $\mathscr{L}_K$, is a many-one function from the expressions in Σ^{**} to the family $\mathscr{P}(\Omega)$ of all the subsets of Ω (i.e. Δ assigns every complete utterance, e.g., a phrase, a certain class of objects).

More explicitly, the membership of Σ is the totality of basic (atomic) signals or letters, including the nonsignal $\Box$. These signals concatenate associatively to form the set of strings (both well formed and ill formed) Σ^*. For example, if a and b are in Σ, then $a \circ (b \circ \Box) = a \circ b$, which may not be the same as $b \circ a$, is a member of Σ^* though not necessarily of Σ^{**}. The maps in ϕ pick those strings (finite concatenations) that count as correct expressions (or well formed formulas) of $\mathscr{L}_K$, i.e. the elements of Σ^{**}. For example, one of the functions in ϕ will be that which carries the couple $\langle (x), Px \rangle$, belonging to $(\Sigma^*)^2$, into the expression $(x)\,Px$, that belongs to Σ^{**}. (Because Σ^* is denumerable, though infinite, so is Σ^{**}. In other words, the language we are describing contains only a countable infinity of sentences. This is a serious limitation but it need not concern us here, as it will have no consequences for our future work.) So far the basic ideas of the *syntax* of a language. (For mathematical linguistics see e.g. Chomsky and Miller (1963), Chomsky (1963), Ginsburg (1966), Marcus (1967), Arbib (1968), or Harris (1968), each with a different standpoint.) These basic ideas allow one to define a number of useful derivative notions, among others the following two.

DEFINITION 1.2 Let Σ^* be the set of strings of a language $\mathscr{L}_K$. If x and y are in Σ^*, then x is a *part* of y iff there are two strings w and z in Σ^* such that $w \circ x \circ z = y$.

This part-whole relation is reflexive, antisymmetric, and transitive as it should be. Should x be an initial (or final) segment of y, then we should take w (or z) to be the blank $\Box$. We shall need this concept in Sec. 2.1.

DEFINITION 1.3 Let Σ^* be the set of strings of a symbolic language $\mathscr{L}_K$. The *length* of a member x of Σ^* is the number of basic signs (counting repetitions) that are a part of x.

(Note the difference between the length of an expression and the complexity of the corresponding construct if any. Any given construct may be expressed by a number of strings of varying length. More on the difference between constructs and their linguistic wrappings in Sec. 1.2.)

So much, or rather little, for the syntax of $\mathscr{L}_K$. Its *semantics* is given by adding to its syntax a set Ω of objects and a coding function Δ. This function associates every expression in Σ^{**} with a collection of objects (perhaps just a singleton) called the *denotatum* of the corresponding expression. For example, in arithmetic we have, *inter alia*, Δ: Numerals$\rightarrow$ Numbers. The function Δ is many-one: any collection of objects may be called by different names. The objects in question may be physical (occasionally other signs) or conceptual. A particular language is specified by determining not only its syntax (essentially its vocabulary Σ and its grammar $\gamma = \langle \phi, \tau \rangle$) but also its semantics – essentially its denotata Ω and its coding function Δ. While the former decrees what strings are well formed, the latter decides which are well informed.

Caution: the *semantics of a language* is sometimes mistaken for a semantic *theory* (e.g. by Katz and Fodor, 1963). The former is an integral part of a language: it consists roughly of its dictionary, i.e. of the coding function $\Delta : \Sigma^{**} \rightarrow \mathscr{P}(\Omega)$. On the other hand a semantic *theory* is a hypothetico-deductive system couched in some language and concerned with explicating semantic concepts. In other words, while the semantics of a language boils down to its dictionary, a semantic theory may include a theory about dictionaries but is not supposed to include any particular dictionary. Likewise a theory of chemical binding does not contain a list of chemical compounds but enables one to explain and predict their formation. (For further criticisms of the Katz-Fodor view see Bar-Hillel (1970).) Enough of this because there are no viable semantic theories of natural languages anyway.

It has become fashionable to include, in the semantics of a language, the conditions under which certain expressions of the language (its sentences) are true, i.e., their *truth conditions*. We shall not adopt this policy for the following reasons. First, because we shall assign truth values to (some) propositions rather than to their linguistic expressions (sentences). Second and more important, because the proposing and discussing of truth conditions or criteria is a task for the special sciences not for a general theory of language. A language should be rich and neutral enough to allow one to express in it any number of mutually incompatible truth conditions. What semantics can do is to study the general concept of a truth condition – provided it does it in the light of the experience gained in the sciences instead of legislating a priori. (More on this in Ch. 8, Sec. 2.4.) So much for the semantics of a language.

A more complete specification of any language $\mathscr{L}_K$ should include its *pragmatics*. The pragmatics of $\mathscr{L}_K$ may be construed as a map Π_K of the set Σ^* of expressions (both sound and lame) into a set of behavior items of the members of K, the users of $\mathscr{L}_K$. But since the effective determination of such a function Π_K is a matter for empirical investigation we shall not be concerned with pragmatics here. (True, there have been attempts to build a priori systems of pragmatics, by abstracting from concrete linguistic circumstances – in particular from the psychical and social make-up of the language users. But such attempts are methodologically as misguided as any other aprioristic approach to empirical matters.) Another reason for leaving pragmatics aside is that, *pace* Putnam (1970), pragmatics presupposes semantics in the sense that before attempting to investigate what person x means by expression y, or how person z uses the truth concept, one should be reasonably clear about the semantic concepts of meaning and truth – just as the physicist makes sure he understands the concept of atomic weight before proceeding to measuring atomic weights. More on this in Ch. 10, Sec. 3.4.

We conclude here our hasty characterization of the concept of a language. We take language for granted and leave it in competent hands – those of linguists, psycholinguists, sociolinguists and historical linguists. Our concern is with what can be said with languages of a special kind, namely the conceptual ones. That is, we are interested in following the Δ arrow.

1.2. *Construct*

From now on we shall restrict our attention to *conceptual languages*, which are the ones employed in mathematics, science, and philosophy. In anticipation of the formal definition to be given in Sec. 2.2 we may say that what characterizes a conceptual language is that some of its expressions symbolize ideas. If we abstract from ideation, which is a concrete brain process, as well as from communication, which is a concrete physical process, we get *constructs*: concepts (in particular predicates), propositions, and bodies of such – for example, theories. Unlike cognitive psychology, psycholinguistics, and pragmatics, all of which are concerned with real people engaged in thinking and communicating, philosophical semantics abstracts from people, and hence does not handle communication. (Nor does mathematical linguistics for that matter.) Philosophical semantics handles constructs *as if* they were autonomous – i.e., Platonic Ideas – without however assuming that there *are* such.

The existence of constructs may be regarded as a pretense on a par with the infinite plane wave and the self sufficient steppenwolf: all three are fictions. (See Vaihinger, 1920; Henkin, 1953.) In each case the real thing is far more complex but, if we want to theorize, we must start by building more or less sketchy models: once the modeling is under way we may contemplate complication and articulation. In any event we shall agree that a conceptual language is a language suitable for expressing constructs of some kind, e.g., biological theories. Moreover we assume that the chief linguistic categories *correspond* to (but are not identical with) the conceptual categories: (some) terms correspond to concepts, (some) sentences to propositions, certain fragments of languages to theories. (Cf. Kneale, 1972.) See Table 1.1., which summarizes the preceding informal characterization of the sign-construct relations.

(Note that we make no use of the notion of a semantic category [*Bedeutungskategorie*] introduced by E. Husserl and worked out by S.

Leśniewski and K. Ajdukiewicz. It does not improve things for us because a semantic category, such as a name or a sentence, is supposed to be "defined by its meaning" (Ajdukiewicz, 1935), and none of those authors had a theory of meaning to offer. In any case their semantic categories are actually syntactic categories, and the elucidation of this concept is a problem for theoretical linguistics not for philosophical semantics. Here we take languages and linguistics for granted.)

TABLE 1.1
Linguistic categories and conceptual categories

	Linguistic category	Conceptual category	
		Variable	Constant
Term	Individual name	Unspecified individual x. Paradigm: arbitrary member of a set.	Specific individual c, or individual concept. Paradigm: "3"
Term	Class name	Unspecified set X. Paradigm: abstract set.	Specific ("concrete") set. Ex.: the set of electrons.
Term	Predicate letter	Unspecified predicate X. Paradigm: arbitrary predicate.	Specific predicate P. Ex.: a concept of pressure.
Expression	Phrase	Definite description schema. Ex.: "The cube of n".	Definite description. Ex.: "The cube of 2".
Expression	Sentence	Statement schema (propositional function). Paradigms: Px, Xa, Xy.	Specific statement. Ex.: "Wolves do not wage wars".
Expression	Language	Abstract theory. Paradigm: Boolean algebra.	Specific theory (= theory of a model). Paradigm: arithmetic.

The unit construct is the *concept*. The concept of a concept cannot be defined without circularity but it may be characterized in various ways. From a logical point of view concepts are the building blocks of statements. (Or the other way around: propositions are certain compounds of concepts.) *Example 1* All three components of the statement $\ulcorner 3 > 2 \urcorner$ are concepts. *Example 2* Every one of the components (coordinates) of the relational structure "$\langle N, +, 0 \rangle$", where 'N' designates the set of natural numbers, is a concept and so is the structure as a whole. *Example 3* Every symbol in Newton's formula for the gravitational force designates a concept, and the formula as a whole designates a proposition (which in turn represents an objective stable pattern).

We may distinguish two kinds of concept: individual, such as "Mars", and collective, such as "planet". The latter, i.e. class concepts, are usually called *predicates*. Predicates can be unary like "long", binary like "longer than", ternary like that occurring in "*a* is *b* times longer than *c*", and so on. Let us for a moment focus our attention upon predicates.

1.3. *Predicate*

We shall analyze the concept of a predicate with the help of the mathematical concept of a function. A *function* f is a correspondence between two sets A and B such that, to every member x of A, there is a single element y of B. The correspondence is written '$f:A \to B$', where A is called the *domain* and B the *range* of f. The value f takes at $x \in A$ is designated by $f(x)$, in turn an element y of B. That is, $f(x)=y$. This is the gist of the general concept of a function.

We are interested in a particular kind of functions, namely propositional functions. A *propositional function* P is a function whose values are propositions. That is, a propositional function, or *predicate*, is a function from individuals to statements. Thus "lives" ("is alive") may be regarded as a certain mapping L from a set D of objects such that, for an individual c in D, $L(c)$ is the proposition "c is alive". Briefly, $L:D \to S$, where the domain D is in this case the set of organisms and S a certain set of statements, namely the class of propositions in which the concept L occurs. Likewise "dissolves" may be analyzed as a function from the set of ordered pairs ⟨solvent, solute⟩ to a set of statements. In general, a predicate (or propositional function) of *rank* (or *order*) n, where n is some integer greater than zero, will be analyzed as a function

$$P:A_1 \times A_2 \times \cdots \times A_n \to S$$

where each A_i, for $1 \leqslant i \leqslant n$, is a set of objects, S is a class of statements (propositions), and the cross represents the cartesian product of the sets of objects concerned. In simpler terms: a predicate of rank n is that which combines n objects, say $x_1, x_2, \ldots, x_n$, not necessarily real nor necessarily distinct from one another, to produce something else, namely $Px_1x_2 \ldots x_n$, called a *statement* or *proposition*, and constituting the value of P at the point $\langle x_1, x_2, \ldots, x_n \rangle$. Moreover, given an arbitrary function $f: A \rightarrow B$ the corresponding predicate will be taken to be the propositional function

$$P: A \times B \rightarrow S \quad \text{such that} \quad Pxy = (f(x) = y) \quad \text{for} \quad x \in A, y \in B.$$

The previous analysis of a predicate as a function applies to atomic, i.e. logically simple predicates, such as "between". Because we have construed atomic predicates as functions we must build complex (molecular) predicates by respecting the formation rules of complex functions out of simpler functions. Thus just as the sum and the product of two functions are definable only on their common domain, so *the disjunction and the conjunction of two predicates must be defined on their overlap* – provided the latter is not void. Otherwise the predicate letter will fail to stand for a genuine predicate: it will be a nonsense sign. In short, we stipulate that, if P and Q are predicates with a common domain $D = A_1 \times A_2 \times \cdots \times A_n$, then

$$\neg P: D \rightarrow S \quad \text{such that} \quad (\neg P)\, x_1 x_2 \ldots x_n = \neg(P x_1 x_2 \ldots x_n)$$
$$P \circ Q: D \rightarrow S \quad \text{such that} \quad (P \circ Q)\, x_1 x_2 \ldots x_n = = P x_1 x_2 \ldots x_n \circ Q x_1 x_2 \ldots x_n$$

where $x_i \in A_i$ for $1 \leqslant i \leqslant n$ and $\circ$ is an arbitrary binary connective, such as "&" or "⇔". The advantages of this construal are multiple. First, it holds for all predicates. Second, it exhibits the referents of a statement. Third, it does not require the concept of truth – whence it is independent of any particular theory of truth. Fourth, it writes off without further ado, as ill formed, compounds such as 'black thought' and 'cleverly melting at 100 °K' because their components, though *bona fide* predicates, are defined on disjoint domains. Such pseudopredicates can be assigned neither sense nor reference.

Note that our construal of a predicate differs from Frege's analysis as a function from individuals to truth values. (Recall Frege (1891), in Angelelli (1967) p. 133: "*ein Begriff ist eine Funktion, deren Wert immer ein Wahrheitswert ist*".) To put it in contemporary mathematical jargon, Frege identifies a predicate F with the characteristic function χ_D of the domain D of F. In short he sets $F = \chi_D : D \rightarrow \{0, 1\}$. But then he is unable to distinguish among the various predicates with the same domain, because there is only one characteristic function for each set. Frege's construal of a predicate is therefore unacceptable. Besides, it is inconsistent with his own (wavering) anti-extensionalism and it involves an unanalyzed concept of truth. (More in Ch. 8, Sec. 3.6.)

On the other hand we accept Frege's conception of a proposition (which he often called *Gedanke*, i.e. thought), namely as the designatum of a declarative sentence independent of its particular wording. (This is a rough characterization not a definition.) That was also the manner Bolzano (1837), the Russell of *PM*, and Church thought of propositions as distinct from their linguistic containers (see, e.g. Church, 1956). We shall not define the concept of a proposition (or statement) but rather the whole structure of a metric Boolean algebra of propositions – not until Ch. 8, Sec. 3.2, though. For the time being we shall clarify what we do *not* mean by a predicate and its values (propositions).

The vain attempt to frame quick definitions of "concept" and "proposition" has produced a number of more or less interesting mistakes. *First:* "A concept is the designatum of a grammatical predicate or predicate letter". Counterexamples: the artificial predicate signs 'crittle' and 'analytic or hot' stand for no concepts. Only the converse is true if qualified: every genuine concept can be either named or described by at least one string in some language. *Second:* "A concept is anything that has a sense (*Begriffsinhalt*)". Or, which is nearly equivalent: "anything which is capable of being the sense of a name *x* is called a *concept of x*" (Church, 1951, p. 11). Unnecessary: there are unnamed (though describable) concepts – witness the silent majority of real numbers and functions. And insufficient: propositions and theories too have a sense. *Third*: "A concept is anything that can be assigned a referent (*Begriffsgegenstand*)". Not necessarily so: sets are concepts and yet they do not refer to anything. *Fourth*: "A proposition is the designatum of a sentence". Close but not quite: the nonsense sentences in nursery rhymes and in some philosophical writings express no propositions. *Fifth*: "A proposition is whatever is either true or false". Not quite: an untested factual statement has no truth value (see Ch. 8). To say that it has got one, only we ignore which, is sheer Platonizing and it does not advance us: the truth of the matter of factual truth is that truth values are contingent upon tests. (More in Ch. 8.) *Sixth*: "A proposition is anything that satisfies the propositional calculus". Necessary but not sufficient: other objects besides propositions obey the same algebra. *Seventh*: "A proposition is a collection of synonymous sentences in a language (i.e., an equivalence class of sentences under the relation of synonymy in a language)". Tempting but hollow, for it does not seem possible to give a purely linguistic account of synonymy. In fact we recognize two sentences as having the same significance, over and above their linguistic disparities, just in case they happen to stand for the same proposition. Example: the different sentences 'p & q' and 'q & p' designate the same proposition. In other words equisignificance, a linguistic property, ir reducible to a semantic property, namely the identity of the underlying constructs. (More precisely, as will be argued in Ch. 7, two constructs are identical just in case they have the same meaning, which happens if and only if they have the same sense and point to the same referents.

So much for a preliminary clarification of "concept" and "proposition". We turn next to theory in relation to language

1.4. *Theory and Language*

The third category of construct is the theory, or hypothetico-deductive system. The formulas of a theory may be specific statements or they may contain unbound variables of some kind. In either case the formulas are formulated in, or expressed by, sentences in some language – the *language of the theory*. One and the same language may be used to formulate any number of alternative theories: when formulating a theory we pick a subset of all the possible expressions of a language and organize this collection (or rather the corresponding set of statements) in a deductive manner. The language itself must be neutral with respect to both the selection and the organization of the material (recall Sec. 1.1). Thus while the language may contain the sentences 'a is a P' and 'a is not a P', the theory will pick either or none of them.

The following example should bear out our thesis that theory and language must be kept as distinct as gift and wrapping. Let $\mathscr{L}$ be the following language:

Logical alphabet of $\mathscr{L} = \{\neg, \vee, \exists\}$

Extralogical alphabet of $\mathscr{L} = \{a, b, x; P\}$ with a and b individual constants, x and individual variable, and P a unary predicate letter

Atomic sentences of $\mathscr{L} = \{Pa, Pb\}$

Sentences of $\mathscr{L} = \{Pa, Pb, \neg Pa, \neg Pb, Pa \vee Pb, \neg Pa \vee Pb, Pa \vee \neg Pb, \neg Pa \vee \neg Pb, (\exists x) Px, \neg(\exists x) Px, (\exists x) \neg Px, \neg(\exists x) \neg Px, \ldots\}$

Theory 1		*Theory* 2
Language of T_1	$=\mathscr{L}=$	Language of T_2
Logic of T_1 = First order predicate calculus = Logic of T_2		
Axiom 1 $\neg Pa \vee Pb.$		Axiom $\neg(Pa \vee Pb)$
Axiom 2 Pa		Theorem 1 $\neg Pa \wedge \neg Pb.$
Theorem 1 Pb		Theorem 2 $\neg Pa$
Theorem 2 $(\exists x)\, Px$		Theorem 3 $\neg Pb$
		Theorem 4 $(\exists x) \neg Px.$

Although the set of sentences of each theory is a subset of the collection of sentences of their common language, the two subsets do not coincide and moreover they do not stand in their own right but "say" something even if this something is abstract for lack of interpretation of the various symbols involved. Whether a system of signs constitutes a *language* (and is consequently neutral) or *expresses* a theory (and is consequently committed) can be decided only by finding out whether it makes a choice among all the possible sentences – i.e. whether it excludes some of the formulas of the given language.

Nevertheless the language/theory division, though genuine, is relative. Indeed every universal theory, whether in logic or in pure mathematics, can be *used as a language* by another, more specific theory. Thus logic is used as a language or vehicle of communication by every mathematical theory, and mathematics is a language of theoretical science. There are two ways in which a theory can be utilized as a language by another theory. One is to borrow only some of the *concepts* of the universal theory and their corresponding symbols without using any of the axioms and theorems of that theory. This is how most mathematicians use logic. It is also the way molecular biologists use information theory: although they talk of the information carried by a DNA molecule they never compute or measure that quantity of information. Another, fuller, utilization of a theory by a second theory occurs when the latter employs some of the *statements* (hence also some of the concepts) of the former. This is how physicists use functional analysis and sociologists graph theory: these theories enter into the very building of the specific factual theories. In sum, the distinction between language and theory, though neat, is as relative as the one between means and goals.

However a relative difference *is* a difference. Hence it is mistaken to define a theory as a certain set of *sentences* from some language, and a model as a structure in which such strings, conceived as *inscriptions*, are satisfied. Such an elimination of constructs in favor of their linguistic embodiments is no mere carelessness but deliberate: it is a necessary component of nominalism. This philosophy, espoused by eminent logicians such as Hilbert and Tarski, and at one time Quine, has its advantages: it simplifies, it avoids the pitfalls of both Platonism and psychologism and last, but not least, any computer can be talked into adopting it.

However, nominalism does not supply an adequate semantic theory. For one thing it overrates the importance of notation and wording and therefore cannot account for the fact that any given proposition can be formulated in a number of different languages. For another, if one refuses to accept anything beyond signs *qua* physical objects and their eventual physical denotata, then he cannot explain why symbols enter into non-physical (e.g. logical) relations – not to speak of their form and content, which are likewise non-physical. (Consider: whereas twice 3 is 6, twice '3' is '33'.) Thirdly, for the same reason the Ockhamist is likely to multiply the number of entities without necessity, as he may take seriously Eddington's parable of the two tables: the layman's (concept of a) smooth table and the scientist's (concept of a) mostly hollow one, both denoted by a single word. The nominalist may thus end up by positing a plurality of realities, thus committing philosophical suicide – as Chwistek (1949) did. Fourthly, if the nominalist assigns to some of his marks, e.g. sentences, certain non-physical properties such as truth or falsity (as Tarski does), then he slips into hylemorphism – the very Platonic devil he wishes to exorcise. Fifthly, the nominalist must repudiate any theory which, like non-standard analysis, teems with nameless constructs (Robinson, 1966). (For further criticisms see Frege (1893) in Geach and Black, Eds. (1952), and Putnam (1971).)

We shall consequently stress the traditional distinction between term and concept, predicate letter and predicate, sentence and proposition, and language and theory. We shall say that the first member of each pair stands for, expresses, or designates the second member. But this relation of designation deserves a separate section.

2. DESIGNATION

2.1. *Name*

Names are those terms in a language that designate objects of some kind. Thus the numerals '3' and 'III' name the number three. In a conceptual language all names designate constructs. But the converse is false: notwithstanding nominalism, most constructs go unnamed. Thus, only a few irrational numbers and a few functions have got standard names. Still, just as anonymous persons can be identified by their features and actions, so constructs can be identified by their properties even if they are assigned no individual names. For example, the function "two fifths of the cube plus seven" has got no special name but it is unambiguously characterized by the preceding description, which amounts to saying what the function does – namely sending x into $\frac{2}{5}x^3+7$, where x is a real number. In short, there are unnamed constructs but not ineffable ones.

There are names of (some) individuals, names of (some) classes (or common names), of (some) relations, and so on. In either case names may designate a fixed or an arbitrary member of a collection, i.e. an unspecified individual. Thus we may agree to call 'x' an arbitrary real number. (Note that a variable is not quite the same as a blank. While all blanks or voids are the same, namely nothing, variables can differ from one another and can be manipulated. Thus '$x+y=z$' is not the same as ' + = '. Nor is a variable something that varies.)

Names are symbols and as such they act as proxies for their nominata. We should not forget what they stand for – unless we happen to be computers. For example, strictly speaking we should not say 'Let R *be* the real line' but rather 'Let R *name* (*designate*) the real line'. However, as long as we are alert to the distinction between signs and their designata we can afford to confuse them in our speech and writing for the sake of brevity. In short, we can indulge in self-designation provided we keep in mind that symbols are just that – conventional signs proxying for something else.

The refusal to distinguish symbols from what they symbolize can be made into a philosophy. It is, indeed, the core of the nominalist or vulgar materialist philosophy of mathematics. This philosophy appeals to the haters of intangibles and the lovers of simplicity: instead of having symbols on the one hand and constructs on the other it offers a single bag of tangible entities, some natural and others (the signs) artificial. Thus a member of this persuasion will claim, for example, that "The expression or string consisting of '(x) ('followed by 'P' followed by '$x \Rightarrow$' followed by 'P' followed by 'x)' is an analytic sentence of the language L". And he will be happy with the nursery school identification of the number one with a vertical stroke. A simple creed indeed – hence one inadequate to tackle the complex facts of life. For example the creed makes no room for the principle that symbols, in particular names, are replaceable because conventional: that "the real thing" is the designatum not its name: that no particular sign is indispensable. (For a vigorous defense of this thesis see Frege (1895). Even Bourbaki (1970, Ch. I), despite its nominalist flare-ups, warns against confusing symbol with designatum.)

The preceding semantic principle, that although names may be necessary (or convenient) none of them is indispensable, has been credited to Shakespeare. In fact, in *Romeo and Juliet* he held that the scent of a rose is name-invariant. This principle, sometimes called the "principle of reference" (Linsky, 1967), can be given a more exact though far less poetic formulation, such as the following. Let x be part of an expression $e(x)$ in a language $\mathscr{L}$ and let $e(y)$ be the expression resulting from trading x for y in $e(x)$, where y is yet another sign in $\mathscr{L}$. If x and y have the same designatum, then so have $e(x)$ and $e(y)$. Consequently the two can be mutually substituted *salva significatione et salva veritate*. One may adopt this necessary triviality as part of the definition of the designation relation.

Thus we accept Shakespeare's thesis that nothing is in a name: that what really matters is the nominatum. If the latter happens to be a construct, not a physical object, then we may say, following Frege and Church, that the name points to the sense *or* to the referent of the construct. But we may as well substitute 'and' for 'or', as the concept of ambiguity has a pragmatic ring about it. In our theory of meaning a sign that stands for a concept performs both functions: it signifies the sense as well as the referent of the construct it designates. (See Ch. 7, Sec. 1.) Thus the class name *Homo sapiens* symbolizes the technical concept of man. The sense of this concept is given by some of the hypotheses occurring in the sciences of man, while its reference class is of course the set of humans. Whether or not this latter set is empty (as some of us are beginning to suspect) is no business of the semanticist: the determination of the actual extension of concepts is a task of scientists not of philosophers. But this matter will be taken up in detail in Ch. 2 and in Ch. 9, Sec. 1.

Names are conventional but they need not be arbitrary. While some are formed spontaneously others are contrived according to rule. About the most sophisticated of all rule-directed naming procedures is Gödel's, which assigns to every basic symbol in the linguistic wrapping of a theory a unique natural number in such a way that, given any natural number, the corresponding symbol can be effectively retrieved. We shall need no such powerful method: the most modest of all naming procedures will suffice us – namely the *quotation marks* technique. We shall employ three kinds of quotes:

> 'single' quotes to mention symbols,
> "double" quotes to designate constructs, and
> ⌜corners⌝ to name propositions (a kind of construct).

2.2. *The Designation Function*

According to Definition 1.1 (Sec. 1.1), a language is a certain septuple $\mathscr{L}_K = \langle \Sigma, \square, \circ, \phi, \tau, \Omega, \Delta \rangle$, where Δ is a correspondence between the signs of $\mathscr{L}_K$ (formed from items in Σ with the help of the operation $\circ$) and the objects in Ω. The role of Δ is then to inject some significance into a bunch of dumb marks, which would otherwise symbolize nothing. In the case of conceptual languages, which are the ones we are interested in, the set Ω shrinks to a subset C, a collection of constructs, and Δ becomes the designation function, briefly $\mathscr{D}$. More precisely, we have

DEFINITION 1.4 Let L be a system of predicate logic. Then the septuple $\mathscr{L}_{KL} = \langle \Sigma, \square, \circ, \phi, \tau, C, \mathscr{D} \rangle$ is a *conceptual language* iff

(i) $\mathscr{L}_{KL}$ is a language;

(ii) Σ contains at least the signs required to designate the basic concepts of the logical system L;

(iii) C is a nonempty set of constructs containing at least all those in L;

(iv) $\mathscr{D}$ is a many-one function from the set Σ^{**} of expressions (wff's) of $\mathscr{L}_{KL}$ into (but not onto) the collection $\mathscr{P}(C)$ of subsets of C;

(v) Any two signs in Σ^{**} with the same designatum (i.e. for which $\mathscr{D}$ takes the same value) can be mutually exchanged wherever they occur in Σ^{**}.

Note the following points. First, since $\mathscr{L}_{KL}$ is a conceptual language, not gibberish, every wff in it stands for (proxies) some construct. On the other hand $\mathscr{D}$ is not onto: there may be in C nameless constructs, such as the nonstandard objects in nonstandard analysis. Second, condition (v), sometimes called 'Leibniz' principle of the substitutivity of identicals', does not warrant any such substitutivety but rather the one of *different* expressions provided they happen to designate the same constructs. Third, this principle has nothing to do with reference. Hence it is not illustrated by ⌜The morning star = The evening star⌝. This last statement is, strictly speaking, false. What is true is that the two descriptions have the same referent – namely Venus. More on this in the next chapter. Fourth, the principle (v) has been challenged because it fails for the so-called "intensional contexts", i.e. in connection with "propositional attitudes" expressed by verbs like 'to know' and 'to believe'. For example, although '3' and 'III' designate the same number, the statement ⌜Everybody knows that 3 = III⌝ is plainly false (see, e.g., Linsky, 1967). We shall not take up this question since it belongs in pragmatics or in psychology (see Ch. 8, Sec. 4.3).

Fifth, the principle (v) is essential to all symbolic thinking. Its explicit statement should contribute to clarifying the way symbols perform their function, which is not to live their own life but to stand for, or deputize for, something else. Thus an equation such as ⌜$a = b$⌝ informs us that the letters 'a' and 'b', though different, name the same object. This trivial remark should help the beginning student of mathematics, who is often baffled by the apparent inconsistency of ⌜$a = b$⌝. And the remark should embarrass the nominalist, for in the formula '$a = b$' the two sides are not the same name. If mathematics were concerned with signs and languages rather than their designata, we should write: 'a' = 'b', which is false. In fact, as Frege (1879) noted long ago, we are here concerned with a single concept, now called 'a', now 'b'. (This does not endorse Frege's view that the theory of identity concerns only names or terms. The theory of identity is supposed to be universal, in the sense that it holds for objects of all kinds.)

Sixth and last, note that we have not defined the designation function $\mathscr{D}$: we have not specified the way it maps expressions into constructs. Any precise characterization of $\mathscr{D}$ involves the exact specification of such a correspondence, therefore of its domain Σ^{**} and range $\mathscr{P}(C)$ – hence a loss of generality. As soon as the designation function is specified, the conceptual language $\mathscr{L}_{KL}$ becomes a particular *semantic system*. A customary way of specifying ("defining") a semantic system is to lay down *designation rules* such as "Let '*M*' designate the Newtonian mass function" (Carnap, 1942). But of course a formula such as this is a rule from a pragmatic point of view only, as it expresses a decision or convention. From a semantic point of view it is not a rule but a statement, namely an instance of "$\mathscr{D}(\sigma)=c$", where $\sigma \in \Sigma^{**}$ is a sign and c the construct σ designates. Besides, designation "rules", though necessary, are not sufficient for the purpose of formulating a body of factual knowledge, however modest it may be. Even an address book involves an additional semantic notion, namely that of *denotation*, a relation which goes from signs to factual items, either directly or via constructs. (More on denotation in Ch. 2, Sec. 2.4.) This relation includes the directory relation that matches names of people to names of places (addresses) and *represents* the physical relation between people and their dwellings:

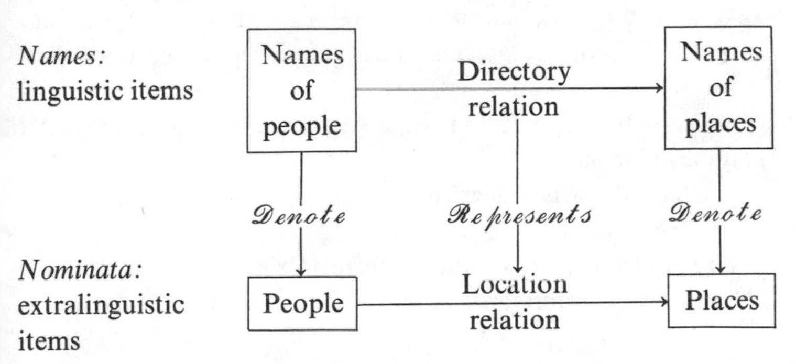

In the case of scientific theories we have the composition of two relations: designation $\mathscr{D}$, from signs to constructs, and reference $\mathscr{R}$, from constructs to factual items. That is, in addition to the ordinary assumptions (construct-construct links) and the designation "rules", scientific theories also contain construct-fact correspondences. The latter tell us what the theories are about: which of its constructs refer to what and which represent what. In other words, the semantic systems of factual science, unlike those of pure mathematics, include semantic hypotheses or assumptions, often misnamed 'correspondence rules' (Chs. 2, 3, and 6). For this reason, and because a language should be a neutral instrument for expressing ideas, our Definition 1.4 of a conceptual language should not be mistaken for a definition of a factual theory. In other words, Definition 1.4 characterizes only a class of languages suitable for expressing a body of factual knowledge.

3. Metaphysical concomitants

3.1. *Basic Ontology*

Our definition of a language in Sec. 1.1 involved not only signs but also their denotata, i.e. the values of the coding function Δ. The denotata may be, but mostly are not, linguistic items. In fact the denotata of a sign system may be anything whatever: individuals, sets, relations, concrete or abstract, possible or impossible. They are *objects* in the general philosophical sense of the word, not in the sense of concrete tangible things.

We shall take this general notion of an object as primitive or undefined, for it is far too basic and important to be definable. And we may let ontology take care of its characterization even though we reject the very notion of a general theory of objects of any kind. In any case we shall assume the following partitions:

(i) Every *object* is either a factual item (e.g. an event) or a construct (e.g. a set) and none is both.

(ii) Every *factual object* is either linguistic or extralinguistic and none is both.

(iii) Every *linguistic object* is either a term (a member of some Σ or of some Σ^*) or an expression (an element of some Σ^{**}) or a whole language.

(iv) Every *construct* is either a predicate or a propositional function or a proposition or a set of either (with or without a structure).

In other words, the basic ontology accompanying our semantics admits the kinds of object shown overleaf.

That there is such a diversity of kinds of object is suggested by the fact that they satisfy radically different sets of laws: for example, the concepts of physics do not obey the same laws as their referents. Whether or not all objects in our ontology are attributed an *autonomous* existence is another matter – one for metaphysics not for semantics. However, the author hastens to declare that in his own metaphysics neither constructs nor even linguistic items are self-existing: the two are artifacts, hence dependent upon man – they were and are being created by mankind and will follow the fate of mankind. Surely signs, e.g. inscriptions, are physical objects: but they depend upon man for their coming into being as well as for their functioning as signs (proxies) for whatever they stand for. As to constructs, they are total fictions: what is real is the brain process that consists in thinking of some object. Let us elaborate on this point.

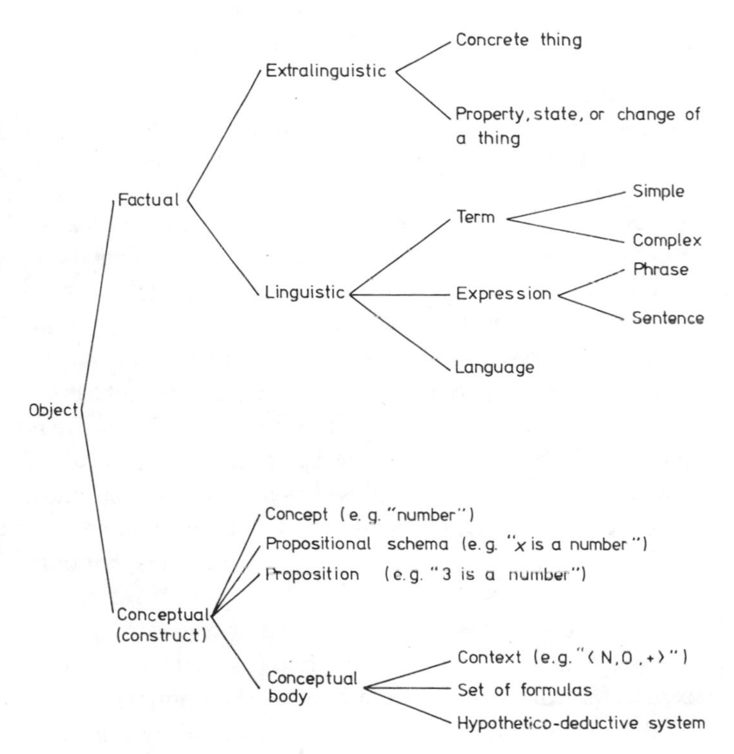

3.2. *Beyond Platonism and Nominalism*

Consider an external event such as a lightning bolt and a linguistic fact such as someone's uttering the sentence 'That was lightning'. Between these two physical processes we have an intermediate brain process – a thought. Change the speaker or the language and the sentence may change. However, we may surmise that, if the speakers' genetic make-ups and their funds of experience are similar, so will be their thought processes, whence their various utterances, though possibly different, will stand for the same statement or proposition. (See Fig. 1.1.)

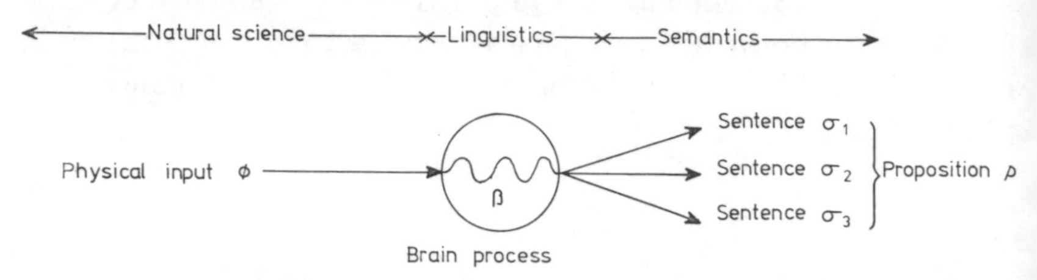

Fig. 1.1. A physical stimulus ϕ, the brain process β (thinking) it elicits, and its linguistic outputs σ_i, sentences expressing the proposition p.

Propositions are not physical objects: they have no reality aside from brain processes – just as there is no motion independent of moving things. It is a fiction, albeit an indispensable not an idle one, to assume that, in addition to the factual items and independently of them, there is such a thing as a proposition. For Platonists like Bolzano there are propositions *in themselves*, that need not have been thought by anyone, hence that may never be "discovered" (Bolzano, 1837). We make no such metaphysical assumption: we make instead the *methodological pretense*, or useful fiction, that things happen *as though* there were autonomously existing propositions (statements) as the designata of some sentences. (Cf. Sec. 1.2.) The concept of existence involved herein is that of conceptual existence not physical existence (Ch. 10, Sec. 4.1). We can turn the Platonist and the fictionalist loose in the field of constructs precisely because these *are* fictions. It is only where real (concrete, material) entities are concerned that the restrain advocated by the nominalist is in place.

Let us take another look at the situation depicted by Figure 1.1. The real events are the lightning, its perception, the thought process triggered in the subject by that perception, and his linguistic utterances. Each of these real events can be studied by some factual science. In particular, linguistics may take care of the sentences – and literary criticism may decree which sentences are "felicitous". Philosophical semantics starts thereafter, where linguistics leaves off. The former is not concerned with speech acts or even with linguistic items for their own sake or as constituents of human behavior but only insofar as they represent constructs.

So much so that philosophical semantics is hardly interested in non-conceptual signs such as 'ouch!' and 'liquid dream' or even in inscriptions, such as hieroglyphs. In other words, semantics is concerned with whatever grows at the tip of a conceptual sign. In this sense semantics is not a part of semiotics conceived as the science of signs (Morris, 1938). Philosophical semantics is a science of constructs and therefore may be regarded as a separate philosophical discipline or else as part of epistemology (see Ch. 10, Sec. 3).

Our view, though not Platonist, contrasts with the nominalist approach to semantics. Whether medieval or contemporary, materialist or empiricist, nominalism trusts the tangible (things and words) as much as it mistrusts the intangible – thoughts and constructs. Language is allegedly ostensible and controllable, whereas thought is hidden and can be wild, and constructs are admittedly fictions. The nominalist deals in terms and sentences not in concepts and propositions (see e.g. Zinov'ev, 1973). The motivations of nominalism are quite sound: to avoid obscurity, wildness, ghostliness, and Platonic hypostatizations. But the surgery it counsels, in particular the excision of constructs, eliminates the semantic problem instead of solving it. The "final solution", consisting in construing everything conceptual as a matter of signs, is in the same vein as the behaviorist proposal of regarding thinking as imperceptible laryngeal movements or as a disposition to talk: the former beheads semantics just as the latter decerebrates psychology. Whatever we may think of nominalist metaphysics, we cannot subscribe to nominalist semantics if we want to understand (*a*) that constructs are not entities on a par with things, (*b*) the very idea of a conceptual symbol as distinct from a nonconceptual one, (*c*) the dispensable character of every single symbol (but of course not of symbolization), (*d*) the fact that conceptual symbols must be adapted to the laws (logical, mathematical or philosophical) of constructs rather than the other way around, and (*e*) that these laws are not factual and are not subject to empirical tests.

The fruitful stand with respect to intangibles is not writing them off but to investigate whether they are lawful and whether positing them explains anything. Thus instead of outlawing the concept of a proposition, logicians have built propositional calculi. Likewise instead of ostracizing the concept of meaning for being intangible, we should clarify it by constructing a theory of meaning, just as scientists have been taming their own intangibles – fields, hydrogen bonds, gene pools, learning abilities, social structures, and the rest. Even linguists go beyond facts in order to understand them: in fact their mathematical models of language discard speakers and speech acts and they contain high brow constructs. Thus the set Σ^* of strings of a language (Sec. 1.1) is infinite; a phoneme is not a particular sound but an equivalence class of sounds – such as the m's uttered by different speakers; and the deep structure of a phrase is not perceptible either. A mathematical model of a language is not an empirical item but a more or less coarse representation of actual facts, to be tested by confrontation with empirical items. In the case of linguistics the empirical items are the finite samples of a language, such as those produced by Shakespeare, Johnny, and so on. Were the linguist to confine himself to such samples he would find hardly any stable and deep regularities: patterns, whether linguistic or chemical, must be hypothesized. Samples suggest and test a model but do not constitute it. Likewise if the semanticist were to limit himself to words and dictionaries he would never come up with theories of meaning and truth. His business is not to observe and describe speech acts but to analyze and systematize the semantic properties of concepts, propositions, and theories. By so doing he need not hypostatize constructs and meanings: conceptualists need not, nay they should not be Platonists. The conceptualist shares the nominalist's conviction that there are no universals in actual discourse – which is a string of concrete events (Goodman, 1951, p. 288). Only, he cannot see how universals could be dispensed with in *theorizing* about "word events", "illocutionary acts", or any other items.

We conclude with Table 1.2, which lists the main philosophies of the sign-construct relation to be discussed in this book.

TABLE 1.2
Sign and construct: main views

View	Sign	Concept, Proposition	Theory
Platonism	A physical object – the shadow of an idea.	A constituent of the Realm of Ideas, which is self-existent.	A body of ideas. Objects of a mathematical theory = Concepts.
Psychologism	A physical object acting as a proxy for some thought.	A thought of some kind.	A set of thoughts, actual or possible.
Nominalism (materialist or empiricist)	A physical (or alternatively experiential) item that stands in its own right or represents some other item of the same kind.	There is none.	A set of expressions: a part of a language. Objects of a mathematical theory = Signs (marks).
Conceptualist materialism	A physical object that stands for another object (or set of objects) – physical, mental, or abstract.	No constructs aside from mental objects, in turn brain processes. Construct = Equivalence class of brain processes – i.e. neither a concrete individual nor a Platonic Idea. For the sake of expediency pretend constructs exist by themselves.	A set of statements with deductive structure. Objects of mathematical theories = Constructs. Scientific theories have an external reference.

CHAPTER 2

REFERENCE

We propose to study now the semantical concept of reference – the *suppositio* of medieval logicians. This concept occurs in statements such as ⌜She was *referring* to witches⌝, ⌜Ecology *concerns* the organism-environment relationships⌝, ⌜Economics *deals with* the production and circulation of merchandises⌝, and ⌜Political science is *about* political institutions⌝. This semantic concept of reference should be distinguished from the psychological or pragmatic notion of reference involved in ⌜That theory suggests (or makes one think of) x⌝, ⌜This theory is meant to apply to x⌝, and ⌜The intended referent of x is y⌝. This other notion of reference is involved in learning how people actually create, learn, or use ideas, whereas the semantic concept of reference comes up when asking what a statement is about quite apart from the way it is conceived, applied, misapplied, or tested.

It might seem that an investigation of the semantic concept of reference should be trivial and therefore useless: don't we normally know what we are talking about? Unfortunately not: there is often endless argument as to what certain scientific theories are about. Hence we should welcome any semantic doctrine that could be of some help in spotting the genuine referents of a scientific theory. Since no such semantic doctrine seems to be available, we shall have to build one. But before proceeding to this task let a few examples vindicate our claim that the referents of a statement, or of a set of statements, are not always in evidence and that there are no obvious tools for bringing them to light.

1. Motivation

Consider the following cases gleaned from the literature.

Case 1 Several philosophers have maintained that ⌜p is true⌝ is identical to ⌜p⌝, i.e. that the truth concept is redundant. On the other hand Bolzano held that the subject (referent) of ⌜p is true⌝ is p itself, which is not the case with the latter statement, whence the two are different. He even suggested the possibility of iterating this procedure, thus coming up with an infinite ladder of statements, every one of which has the previous one as its subject or referent (Bolzano, 1851, p. 85).

Case 2 The same philosopher had held previously that ⌜Some people are literate⌝ concerns only literate persons while ⌜Some people are not literate⌝ is about illiterates only (Bolzano, 1837, III, Sec. 305). True or false? And how about their equivalents ⌜Not everyone is illiterate⌝ and ⌜Not everyone is literate⌝ respectively?

Case 3 Aristotle taught that "A single science is one whose domain is a single genus" (*Post. Anal.*, Bk. I, Ch. 28). Right or wrong? How about ecology?

Case 4 Evolutionary biologists are divided on the question of the referents ("units") of population genetics and of the theory of evolution (see Williams, 1966, Ch. 4). Is it individual organisms, species, or populations? And does the theory state that selection acts on genotypes (individuals) or rather on phenotypes (populations)?

Case 5 Some proponents of the so-called identity theory argue that, although neurophysiology and psychology employ concepts with different senses, they have exactly the same referent, namely the person, whence those sciences constitute different ways of viewing the same facts, namely mental (=neurophysiological) events. This particular defense of the identity theory rests then on the semantic hypothesis that Referent= =Fact. How about the ethology and the physiology of birds, which share their referents?

Case 6 Eminent physicists have claimed that the special theory of relativity is about the behavior of clocks and yardsticks. Others have held that it concerns observers in relative motion. Still others, that the referents of the theory are point masses inhabited by competent and well equipped experimenters communicating with one another through light signals. Finally, there are those who contend the theory to be about any systems connectible by electromagnetic signals. Take your pick.

Case 7 Most physicists act on the assumption that the quantum theory refers to autonomously existing microsystems such as neutrons and photons. But when it comes to "philosophizing" many claim that the theory is about sealed (unanalyzable) blocks constituted, in arbitrary proportions, by microsystems, measuring instruments, and observers.

Still others hold that the theory is concerned with the knowledge of nature rather than the latter (Heisenberg, 1958, p. 100).

Case 8 Consider the formula ⌜For all integers x: $x+1=1+x$⌝. Is it just about integers or rather about the whole system $\langle Z, 1, +\rangle$ (Rosenbloom, 1950, p. 110)? In any case, by which criterion is it either?

Case 9 The theory of control systems has been developed by engineers but has applications in biology and other fields of inquiry concerned with things, whether inanimate or alive, that have control devices built into them. Does the theory have a definite reference class at all?

Case 10 What is the reference class of elementary logic? Combinations of letters? Sentences? Propositions? Argumentative people? The world?

There should remain little doubt that identifying the referents of a statement or of a theory can be a thorny problem and that we need a semantic theory capable of helping us to perform that task. We shall presently expound one such theory. More precisely, we shall first study the rather unruly relation of reference and shall later on introduce a couple of law-abiding reference functions – one for predicates, the other for statements. The outcome may be regarded as a calculus of reference allowing one to compute the reference class of any composite statement as a function of the reference classes of its constituents. This calculus should help us to solve problems of referential ambiguity. By the same token it should free us from the recourse to authority as the "method" for ascertaining what scientific theories are about. On the other hand our calculus will not presume to tell us when a given predicate is applicable, i.e. what its correct reference or field of validity is: this is a business of science. In other words, we distinguish reference from extension – a subject to be studied in Ch. 9, Secs. 1 and 2.

2. The Reference Relation

2.1. *An Unruly Relation*

We stipulate that the reference relation $\mathscr{R}$ holds between constructs (concepts, statements, or theories) on the one hand, and objects of any kind on the other. In other words, we adopt the following

CONVENTION The graph (or extension) of the reference relation $\mathscr{R}$ is a set of ordered pairs construct-object, i.e.,

$$\mathscr{E}(\mathscr{R}) \subseteq C \times \Omega, \quad \text{with} \quad C \subset \Omega,$$

where 'C' stands for the class of constructs and 'Ω' for the class of objects.

$\mathscr{R}$ has no simple formal properties. In particular $\mathscr{R}$ is not reflexive throughout its graph. For example, the concept of a star concerns stars rather than itself. On the other hand the number 7 refers to nothing. Nor is $\mathscr{R}$ either symmetric or antisymmetric. Finally $\mathscr{R}$ is not transitive either, as shown by the following counterexample:

$$r = \ulcorner\text{The statement below is false}\urcorner \tag{1}$$

$$s \tag{2}$$

where s refers to a third statement t. Clearly, although $\mathscr{R}rs$ and $\mathscr{R}st$, it is not the case that $\mathscr{R}rt$.

This last result has an important application in the foundations and philosophy of science, namely in relation to the semantic status of metanomological statements, or laws of laws (Bunge, 1961a; Angel, 1970). Consider the following propositions:

> A fundamental physical statement (e.g. equation) should involve no constants other than universal constants. (3)
>
> Newton's laws of motion are invariant under Galilei transformations. (4)
>
> Every quantum-mechanical formula should correspond to some classical formula. (5)
>
> If a local field theory is relativistically invariant, then it is also invariant under the combined conceptual inversion of charge, time, and parity. (6)

These statements and many others – some descriptive, some prescriptive – are often treated on a par with the object statements of a theory. However, it is apparent that they are *metastatements*, i.e. that they concern further statements. (Caution: not every metastatement belongs to some metatheory. None of the above does. Moreover some metastatements belong to object theories – as is the case with (4) and (6) above. For a typical confusion of 'metastatement' and 'metatheoretical statement' see Freudenthal (1971).) Further, since the reference relation is not transitive, the above propositions fail to refer to the same object their referents are about. Thus statement (4) is not a law of motion and (6) is not a field law – hence neither can be tested by observations on physical objects. Which suggests that, in spite of operationism, semantics should precede methodology: before we pose the problem of testing a statement we should know what it refers to. Enough of this for the moment.

In conclusion, $\mathscr{R}$ is neither reflexive nor symmetrical nor transitive – nor, indeed, seems to have any other definite formal trait. It is a wishy-washy relation and as such not amenable to theory. All our remarks in this section will therefore be intuitive. To obtain regularity we shall have to introduce a reference function. This will be done in Sec. 3; but before that we must take a closer look at $\mathscr{R}$.

2.2. *Immediate and Mediate Reference*

Consider a specific theory, or theoretical model t, of some concrete system s. Any such specific theory "defines" a model object m, or conceptual image of s, that is hoped to capture some of the traits of the latter. We may say that, while the *immediate referent* of t is the model object m, the ultimate or *mediate referent* of t is the real thing s (Bunge, 1967a). Table 2.1 exemplifies the idea.

TABLE 2.1

Examples of immediate and mediate reference

Theoretical model t or specific theory	Model object m = immediate referent of t.	Real system s = immediate referent of m. = mediate referent of t.
Enterprise theory	Directed graph	Enterprise
Contagion theory	Diffusion equation	Epidemics
Volterra's predation equations	Predator-prey system in constant environment	Rabbit-fox system or the like in wild life
Eye dioptrics theory	System of lenses	Mammalian eye
Magnetostatics of magnetic dipoles	Magnetic dipole	Terrestrial magnetic field

In every such case (*a*) the formulas of the theoretical model (specific theory) concern directly the model object itself, and mediately the thing modeled: for example, a theorem in enterprise theory may be about the degree of a vertex in the hierarchical tree of the enterprise; (b) the formulas are true of the model object (e.g. the unperturbed rabbit-fox economy) but only approximately true of the real thing – e.g. the rabbit-fox system in a variable environment subject to droughts, viruses, etc.; (*c*) the reference relation holds between every pair: $\mathcal{R}tm \& \mathcal{R}ms \& \mathcal{R}ts$ even though it is not transitive.

2.3. *Reference Class*

The set of referents of a given construct *c* is called its reference class. More explicitly, we introduce

DEFINITION 2.1 If *c* is a construct, then the *reference class* of *c* is the set of objects referred to by *c* (or the collection of items to which *c* bears $\mathcal{R}$), i.e. the relation class $[c]$ of *c* relative to $\mathcal{R}$:

If *c* is in *C*, then $[c] =_{\text{df}} \{x \in \Omega \mid \mathcal{R}cx\} = \overleftarrow{\mathcal{R}}`c$.

DEFINITION 2.2 A construct *c* *refers partially* to a class $A \subset \Omega$ iff *A* is included in the reference class of *c*, i.e., if $A \subseteq [c] = \overleftarrow{\mathcal{R}}`c$.

Some constructs concern a single natural kind while others are about heterogeneous classes. (We do not apologize for employing the concept of a natural kind, essential to science, but we leave its elucidation to ontology.) For example, the reference class of "viscous" is the set of fluids, while that of "writing" is composed of the set of people and the set of written symbols. This difference is consecrated in the following

DEFINITION 2.3 A reference class is said to be *homogeneous* iff it is composed of elements of a single natural kind.

DEFINITION 2.4 A nonempty reference class is called *inhomogeneous* iff it is not homogeneous.

Every statement in factual science has a reference class assumed to be nonempty even though, on closer investigation, that class may prove void. For this reason it is often advisable to speak of *hypothetical* (or assumed) reference classes. A hypothetical reference class is often called an *intended* reference. It is not convenient to employ this second term in semantics, as it suggests the psychological concept of intention. (Whether or not somebody intends to entertain the hypothesis that a given construct has a certain reference class is a matter for psychology.) The class of objects for which a statement or a theory holds true may be called its *actual reference class*.

Since a scientific theory is a set of factual statements (with a deductive structure) every such theory has a reference class. This class is known or assumed to be nonempty. If the assumption has not yet been substantiated, then the theory itself may point the way to its referents. For example, a theory concerning a hypothetical extinct biological species will be instrumental in the search for fossil evidence relevant to the hypothesis. Should the hypothesis turn out to be false, the actual reference class of the theory would shrink to nought, but the assumed reference class would remain nonempty though indeterminate and idle, at least until further notice.

The reference class of a theory, whether it be the assumed or the actual one, is sometimes called the *ontology* of the theory. This is a misnomer, for an ontology is not a set of things but a philosophical theory concerning the basic traits of the world. Strictly speaking, the *ontology of a scientific theory* is the set of ontological (metaphysical, cosmological) hypotheses presupposed or countenanced by the theory. (See Bunge, 1973b.) For example, the reference class of classical electrodynamics is constituted by the set of bodies and the set of electromagnetic fields. On the other hand the ontology of the same theory is composed of wide ranging generalizations such as "Every thing is either body or field or a compositum of body and field", "The world is a plenum not a collection of atoms in void", "Except at boundaries all properties are continuous", "All properties are lawfully interrelated", and so on. Alternative systems of electrodynamics may or may not have different reference classes. If a change in scientific theory involves a modified reference class then the associated ontology may or may not change. (For example, action at a distance electrodynamics do not involve the field concept and consequently their ontology is closer to Greek atomism than to Aristotle's or Descartes' plenistic cosmologies. On the other hand the exact number of kinds of elementary particle and their precise properties are irrelevant to metaphysics.) One and the same scientific theory is consistent with a multitude of ontologies – all of a feather, though. Although science determines its own metaphysics, it does so only in outline: there is considerable leeway.

The preceding considerations on the relation between the reference class of a theory and its ontologies does not apply to formal science. The reference classes occurring in logic and mathematics are constituted by conceptual objects. For example, the reference class of the propositional calculus is the set of all propositions; number theory is concerned with numbers; topology is about topological spaces – and so on. In general: in formal science $\mathscr{R}$ never points to anything outside formal science. In other words, within formal science $\mathscr{R}$ pairs off constructs to constructs, i.e. $\mathscr{E}(\mathscr{R}) \subseteq C \times C$. To put it another way, we adopt

THESIS 2.1 *If c is a construct in formal science (logic or mathematics), then* $[c] = \overleftarrow{\mathscr{R}}`c \subseteq C$.

This is the postulate of the formal (nonfactual and a fortiori non-empirical) character of logic and mathematics. (Cf. Kraft, 1970.) If the thesis of the autonomy of formal science is accepted then it becomes clear (*a*) why the semantics of logic and mathematics, i.e. model theory (including Tarski's theory of truth), is hardly relevant to the semantics of factual science; (*b*) why logical and mathematical theories are not tested in the laboratory, and (*c*) why logic and mathematics have no associated ontologies – since ontologies are, by definition, theories about *entia* or existents. (If one insists on speaking of the 'ontology of formal objects' then one ought to identify it with formal science not with a branch of philosophy.) However, all this is controversial and will be taken up again, particularly in Ch. 10, Sec. 4.

2.4. *Factual Reference and Object Variable*

Factual items, such as things and events, are conveniently characterized as nonconceptual. To be more precise, we introduce

DEFINITION 2.5 The totality F of *factual objects* is the subset of Ω formed by nonconceptual elements and such that F is disjoint from C:

$$F =_{df} \{x \in \Omega \mid \neg(x \in C)\}.$$

If a concept, statement, or theory refers to one or more factual objects, it will be said to be *factual*. The subrelation $\mathscr{R}_F$ of factual reference pairs constructs off to factual objects, i.e. its extension is contained in the set of construct-fact pairs:

$$\mathscr{E}(\mathscr{R}_F) \subseteq C \times F.$$

This particular concept of reference comes with a particular concept of reference class as characterized by

DEFINITION 2.6 If c is a construct, then the *factual reference class* of c is the set of factual objects referred to by c:

$$c \in C \Rightarrow [c]_F = \overleftarrow{\mathscr{R}}_F`c =_{df} \{x \in F \mid \mathscr{R}cx\} \subseteq F.$$

Factual science contains both factually referential concepts, like the one of attention, and concepts with an empty factual reference, such as "2". The same holds for metaphysics or ontology. Having a nonempty factual reference class suffices for a construct to belong either in science or in metaphysics. More explicitly, we state

THESIS 2.2 If $c \in C$ and $\emptyset \subset [c]_F \subseteq F$, *then c belongs in factual science or in metaphysics.*

Having a precise mathematical form is no indication of ontological status: a factual concept, i.e. one with factual referents, may have a precise mathematical form while a nonfactual concept may fail to have one. Until one century ago some of the concepts of the infinitesimal calculus were in the latter condition.

Scientific concepts are often partitioned into constants and variables. Many of them, whether of the first kind or of the second, are quantitative concepts, and most of these are assigned dimensions and units. Thus the dimension value of a speed is LT^{-1}, and a standard speed unit is $cm\ sec^{-1}$. But dimensions and units give no precise indication of semantic status: any ratio of magnitudes with the same dimension value is dimensionless, hence unit free, even though it may concern some concrete object or other. And dimensional constants, like the gravitational constant γ, are not the value of a property of some physical system, so that they are devoid of factual reference even though they occur in law statements (Bunge, 1967b). A suitable semantic partition of the specific or technical concepts of factual science is rather the one shown overleaf.

What we have somewhat improperly called *scale concepts*, such as L and *cm*, have no factual reference. They are ingredients of some other concepts which do have a factual reference. Thus speeds, with dimension value LT^{-1} and reckoned on a suitable unit, are always speeds of something. This something, usually hinted at in the context, is of course the referent of "speed" – car, light wave, or what have you. In contrast, proportionality constants and dimensional constants, whether or not universal (independent of any particular material), are factually nonreferential: they concern nothing in particular. Thus unlike the velocity

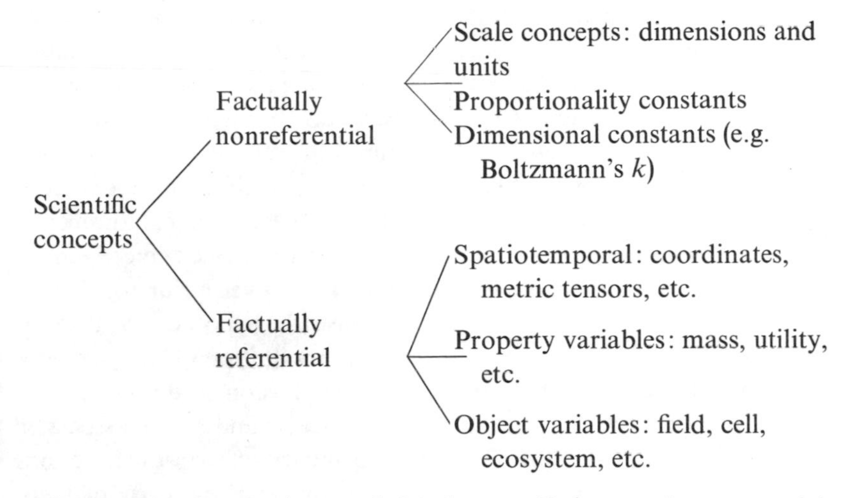

of light or the electric charge *of* the electron, Boltzmann's constant k is not the value of some property of a physical system.

The variables that are factually referential can in turn be classed into: (*a*) *object variables*, referring to things such as photons or persons; (*b*) *property variables*, representing properties of concrete things and relations among them, and (*c*) *spatiotemporal variables*, concerning the basic world framework. Variables of all three kinds are likely to occur in one and the same statement. Examples:

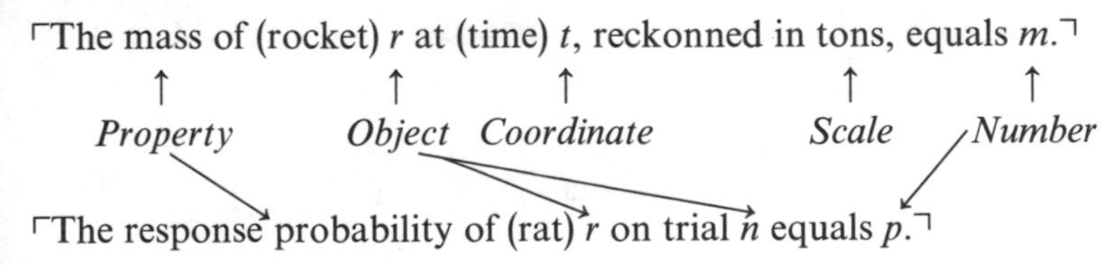

The object variable (or constant, as the case may be) is not usually indicated in an explicit manner, except when a number of individuals (atoms, rats, nations) are involved, in which case each thing is assigned a numeral, a letter, or some other symbol – as is the case with ⌜P_i is the population of the ith country⌝. However, of all the variables the object variables are the most important. Without them the property variables would be baseless and the spatiotemporal variables would be purely mathematical constructs. Indeed, a property other than a purely formal (mathematical) property is a property of some concrete individual or other, whether actual or conjectural. And the spatiotemporal variables occurring in factual science have a similar though less apparent concrete support. Thus in physics every distance value is a distance between two points in some concrete system. And a time interval is the time separation between two events, one of which may be taken as the start of a process. No things, neither properties nor events – nor space or time. However, in the case of spatiotemporal variables it is usually possible and convenient to feign that the spatiotemporal framework is given prior to, and independent of, things and events. This pretense (the hypothesis of absolute space and time), if taken literally, ensues in an illusion of thinglessness, hence of absence of factual reference. The illusion is dispelled by a foundational analysis of spatiotemporal concepts and by the associated metaphysical construction of a relational theory of spacetime, i.e. one according to which space and time are defined on the class of facts. Without such a metaphysical groundwork the semantic analysis of statements containing spatiotemporal variables would be incomplete and misleading. Thus semantics and metaphysics, far from being mutually exclusive, are complementary. (More in Ch. 10, Sec. 4.)

2.5. *Denotation*

Consider the following table.

Linguistic object —$\mathscr{D}$→	*Construct* —$\mathscr{R}$→	*Object*
Proper name	Individual concept	Individual
Predicate letter	Predicate	Property
Declarative sentence	Proposition	Set of facts

The relations $\mathscr{D}$ of designation and $\mathscr{R}$ of reference can be coupled to produce a relation that pairs off signs to objects. This relation, to be designated by Δ, will be construed as the (relational) product of $\mathscr{D}$ and $\mathscr{R}$. That is, we adopt

DEFINITION 2.7 Let Σ^{**} be the set of expressions of a conceptual language. The relation Δ, with domain Σ^{**} and codomain the set $\mathscr{P}(\Omega)$ of classes of objects, and such that $\Delta = \mathscr{D} \cap \mathscr{R}$, is called the *denotation* relation.

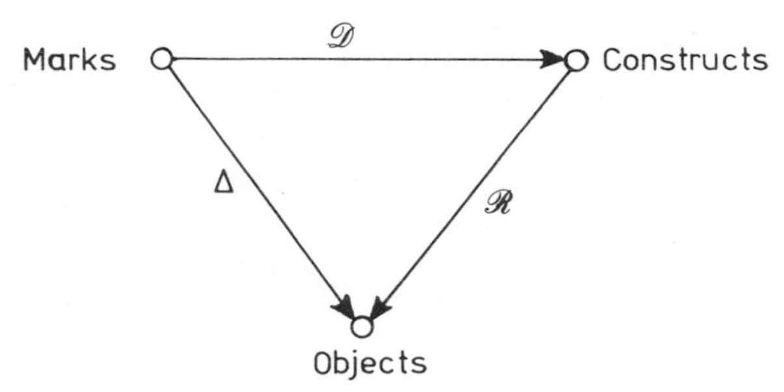

When the denotatum of a sign happens to be conceptual, Δ coincides with $\mathscr{D}$. Thus

'3' designates "3" = '3' denotes "3".

In other words, within formal science there is no difference between designation and denotation. But as soon as a construct refers to some extraconceptual object the difference becomes patent, and with it the impotence of the semantics of formal science to account for factual science. This concept of denotation has the advantage that it could not possibly be mistaken for the notions of reference and extension, as it so often happens with the presystematic notion of denotation (Geach, 1962). We can now use it although we shall not need it often.

2.6. *Reference and Evidence*

Reference classes should be kept distinct from bodies of evidence or evidence classes. Thus a hypothesis concerning the last glaciation period does not refer to the scars left by glaciers on rocks or to the distribution of certain plant species in our day. Likewise a theory about schizophrenia is not to be mistaken for a theory of the behavioral manifestations of it. This sounds obvious and has been pointed out before (Feigl, 1950), yet it has been controverted and is still being denied by die-hard operationists.

In general, the reference class of a scientific theory does not coincide with the body of evidence relevant to it – which may be nil. The two relations involved herein are radically different: whereas the relation of factual reference pairs off constructs to facts or things, the relation of evidence has the *subset* of observable facts as its domain and a subset of the totality of constructs as its codomain. (See Figure 2.1.) In other words, only some facts – namely those accessible to observation – count as evidence for or against a statement that refers to any given class of facts.

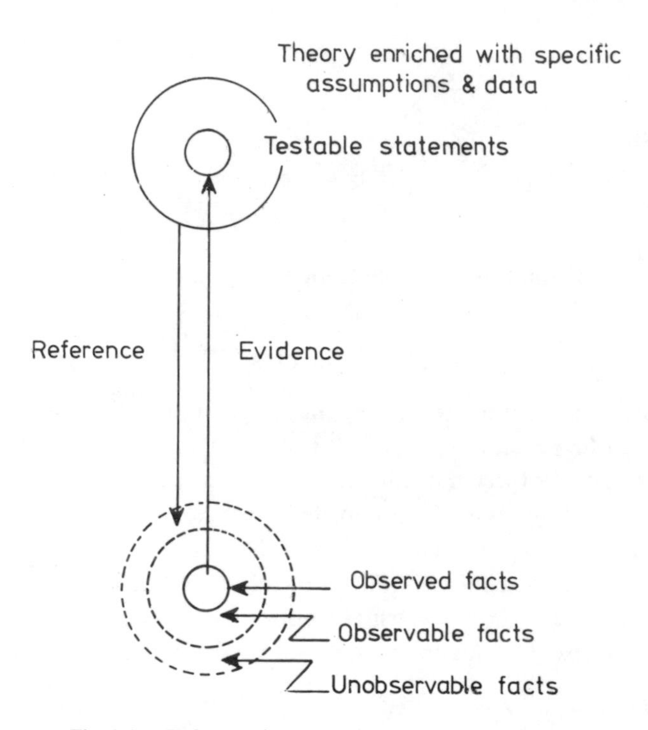

Fig. 2.1. Reference is at best the converse of evidence.

And only some of the statements in a theory, even when adjoined specific assumptions and data, are susceptible to empirical test: all others must remain content with vicarious evidence, if any. (For details see Bunge, 1967a, Ch. 5 Sec. 5.6; Ch. 8 Sec. 8.4; Ch. 12; 1973a, Ch. 2; and 1973b, Ch. 10.) In sum *Reference ≠ Evidence*. Consequently *Reference class ≠ ≠ Evidence class*.

The differences between reference and evidence are best understood in relation to any theory concerning entities that are not accessible to direct observation, such as microphysical theories and economic theories. In this case the testability of the theory depends not so much on the size and kind of its reference class as on the existence of auxiliary theories capable of bridging the gap between the hypothetical referents and their observable concomitants, as well as on the experimental techniques capable of switching on and recording such unobservable-observable couplings.

The size of the reference class of a theory is irrelevant to the testability of the theory. Thus a theory explaining a unique process, such as the formation of our Moon, or the crumbling down of the American Empire, can have a sizable body of evidence. On the other hand a theory about an extremely large number of individuals, such as the neutrinos, is bound to have a comparatively poor evidence class due to the weakness of the interactions among neutrinos and other things. In short, the size of the reference class of a scientific theory furnishes no indication whatever about its testability. Conversely, the size of the evidence class of a theory is of little help in determining the reference class.

Consequently it is unnecessary, nay impossible, to elaborate on the testability-reference relation. All we can do is to issue warning notices and to enunciate the following

THESIS 2.3 If a theory is empirically testable then it has a nonempty hypothetical reference class.

The converse is false: hypothesized reference does not ensure testability. (Thus we can imagine, and theorize on, things that are in principle inscrutable, like the soul of plants.) Nonempty reference is then necessary for empirical testability. But it is insufficient. Consequently before we inquire into the possibility of testing a theory we must find out what it refers to. (Imagine trying to design an experiment for testing a hypothesis with no definite reference, such as a conjecture about a possible world.) Consequence for philosophy: *The semantics of science should precede the methodology of science.* This conclusion contradicts the so-called verifiability doctrine of meaning, according to which the meaning (and in particular the reference) of a statement consists in the way it is verified or, more generally, put to the test. This doctrine has constituted a stumbling block in the development of semantics, for it has mixed up questions of semantics, such as those of sense and reference, with problems of methodology, such as that of empirical testability. (Worse: it has left the latter unanalyzed.) Positivist semantics has had its chance and has failed. Give now realist semantics a chance.

2.7. *Misleading Cues in the Search for Referents*

Before taking the trouble to build a theory of reference we should check whether any of the existing views on the matter might help us to spot the referents of any given scientific formula. Let us therefore make a quick – hence necessarily somewhat unfair – review of the most popular opinions on the subject.

(i) *Grammaticalism:* "The denotatum of a sentence is always denoted by the grammatical subject". This view is false if grammatical predicates are regarded as unary, partially true if their actual complexity is taken into account. Thus traditional grammar would tell us that the referent of ⌜Woman is sweeter than man⌝ is the fair sex, while in point of fact the statement refers to both sexes. A suitable enlargement of the notions of grammatical predicate and subject might help us out in the case of ordinary language expressions. But it would be insufficient for spotting the referents of a statement clad in mathematical language: grammar cannot and does not want to attain the fine structure accessible to mathematics. It would be pointless to ask the grammarian to find the reference class of, say, a Gaussian distribution or a differential equation. Only an analysis of the whole theory in which the formula occurs could produce the desired result. And such an analysis, which lies within the scope of the semantics of science, is beyond the scope of grammar – even of the Chomsky type.

(ii) *Psychologism:* "Search for the intentional object, i.e. for the object of consciousness associated with the given construct". This recipe of Brentano's is definitely misleading: one and the same formula can be "read" in different ways by different persons. In other words, the intended referent of a construct depends not only on the construct but also on the person who thinks of it and the circumstances under which he or she is thinking. This is of interest to psychology not to semantics: the latter is concerned with the hypothetical though possibly real referents of constructs.

(iii) *Linguistic pragmatism:* "No expression refers by itself: it is only some user who, under definite circumstances, can attach a referent to an expression. Hence the act of referring, not reference, should be the object of semantic analysis". While it is true that expressions have no reference unless someone uses them, there is nothing wrong with abstracting from users and circumstances to speak of the reference class of a construct as a fixed set. Surely scientific theories would not exist unless there existed beings capable of making and using them. But the objectivity of such theories depends on their being assigned reference classes assumed (rightly or wrongly) to consist of real entities. (The concept of semantic objectivity is involved herein. That of methodological objectivity will concern us in Ch. 10, Sec. 1.2.) Paradoxically enough, there is no objectivity without abstraction from users and circumstances. If someone claims that a certain theory concerns A's rather than B's, he is expected to substantiate this claim by analyzing the initial assumptions of the theory rather than his own speech acts. Reference hypotheses must be impersonal and open to public inspection.

(iv) *Classical operationism:* "The meaning (=denotation) of a piece of science is to be sought in the operations aimed at testing it: a formula is meaningless unless the procedures for verifying it are specified." This thesis boils down to the conflation of reference and evidence, criticized in Sec. 2.6. Besides, the operationist criterion does not help us in spotting the hypothetical referents of a predicate, for predicates are not testable. And it is misleading in relation to statements, for it diverts our attention from reference to evidence: from the real thing (e.g. loving) to its observable manifestations (e.g. blushings), from facts to documents about them. Empirical tests must be preceded by reference analyses: without knowing or assuming what a given statement is about we would not know how to put it to the test.

(v) *Frege:* "Statements denote (*bedeuten*) truth values". While names and definite descriptions can denote objects, the denotation (*Bedeutung*) of a statement is either truth or falsity. Thus whereas the description "2^2" denotes the number four, the statement $\ulcorner 2^2=4 \urcorner$ denotes truth, and so does $\ulcorner 2>1 \urcorner$. Consequently

$$(2^2=4)=(2>1)$$

"is a correct equation" (Frege (1891) in Angelelli, p. 132). The ambiguous use of the term 'denotation' and the absurdity it leads to would deserve no mention were it not that this slip has been blown up into a doctrine that has misled many a contemporary semanticist. Moral: Keep reference and truth value separate. (Another: Honor the great man's hits not his misses.) A statement *has* a reference class, may be *supported* (or undermined) by an evidence class, and it may be *assigned* a truth value.

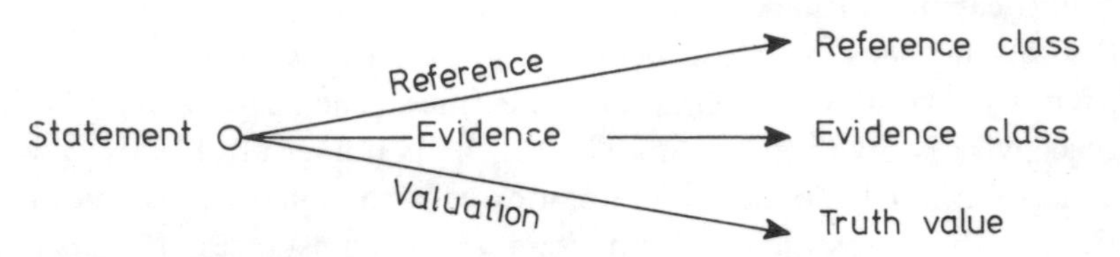

(vi) *Phenomenology:* "Referents are a nuisance: just bracket them out". According to Husserl, to get a direct spiritual vision of an essence (*Wesensschau*) we must discard (*ausklammern*) the external world and drop all previous knowledge of the object. No comment.

Conclusion: The popular views on reference, in particular operationism and Wittgenstein's, do not help us to identify the referents of a construct. Moreover they are misleading. Worse: neither do any of the elaborate and articulate systems of semantics help us out, particularly in relation to scientific constructs. Thus Carnap (1942, 1947), whose work was no doubt exact, elaborate and systematic, did not propose a theory of reference and was hardly interested in our subject. This may have been due to the mistaken impression that reference presents no real problem. Which may be true in the cases of ordinary knowledge and mathematics. But scientific constructs, which dwell between those two covers, do pose a challenge, as we saw in Sec. 1. Let us accept this challenge.

3. The Reference Functions

3.1. *Desiderata*

Consider the table on the next page.

A first point to note is that we are tacitly distinguishing reference from extension. Thus in the 2nd row we state that "Clairvoyance" concerns mind readers and the like without claiming that there are any such

Construct		Referent(s)
1	Human	Mankind
2	Clairvoyance	Mind readers, soothsayers, etc.
3	Conducting	Bodies
4	The moon is round	Moon
5	The moon is round or the cat is fat.	Moon and cat
6	The moon is round and the cat is fat.	Moon and cat
7	All quarks are charged.	Quarks
8	Nobody has so far observed any quark.	Quarks and people
9	Some people loathe animals.	People and animals
10	Nobody loathes animals.	People and animals

individuals. Moreover, if pressed we might grant that the extension of that predicate is nought, while insisting that its assumed reference class is composed of the individuals who claim to have that ability. Likewise in the 7th row we talk about the hypothetical entities called quarks, which at the time of writing have not been identified in the laboratory: the extension of "Quark" is not known. In short, as a rule $\mathscr{R}(c) \neq \mathscr{E}(c)$, where $\mathscr{R}$ and $\mathscr{E}$ designate the reference and the extension functions respectively. In this we follow Buridan (cf. Geach, 1962).

Secondly, it can be observed that reference is rather insensitive to the propositional connectives, i.e. to the gross propositional structure of a statement. Look at lines 5 and 6: the disjunction and the conjunction of two given statements are assigned the same referents. And lines 9 and 10 suggest that a statement and its negate should be assigned the same referents – which is reasonable since they are about the same things. Finally, lines 7 and 8 suggest that a theoretical statement should not be mistaken for a pragmatic one: while the former is supposed to have an objective reference, the latter is bound to concern, at least partly, the cognitive subject.

The above example and others suggest that a theory of reference should satisfy the following desiderata.

*D*1 The theory should define at least one reference function, not just a relation, if it is to enable one to compute the reference class of a complex predicate or formula out of the assumed reference classes of its components.

*D*2 The range of the reference function(s) should be a set of sets rather than the collection Ω of objects, since a single construct may concern a whole class of things.

*D*3 The reference class of a statement and of its negate should be the same.

*D*4 The reference class of an arbitrary propositional compound should equal the union of the reference classes of the components.

*D*5 The reference class of a relation should equal the field of the latter.

The functions to be introduced in the next subsection will prove to satisfy the preceding desiderata.

3.2. *Principles and Definitions*

We need two reference functions, one for predicates, the other for statements. These two types of construct are related in the following way (cf. Ch. 1, Sec. 2.2). An n-ary predicate P is a function

$$P: A_1 \times A_2 \times \cdots \times A_n \rightarrow S$$

from n-tuples of objects (individuals) to statements, such that the value of P at $\langle a_1, a_2, \ldots, a_n \rangle$ in $A_1 \times A_2 \times \cdots \times A_n$ is the atomic statement $Pa_1 a_2 \ldots a_n$ in S. The reference function for predicates will be defined on a set of predicates, and the one for statements will be defined on the set of statements formed with those predicates. The former is introduced by

DEFINITION 2.8 The *reference class of a predicate* is the collection of its arguments. More precisely, let $\mathbb{P}$ be a family of n-ary predicates with domain $A_1 \times A_2 \times \cdots \times A_n$. The function

$$\mathscr{R}_p: \mathbb{P} \rightarrow \mathscr{P}\Big(\bigcup_{1 \leqslant i \leqslant n} A_i\Big)$$

from predicates to the power set of the union of the cartesian factors of the domains of the former is called the *predicate reference function* iff it is defined for every P in $\mathbb{P}$ and

$$\mathscr{R}_p(P) = \bigcup_{1 \leqslant i \leqslant n} A_i .$$

Example 1 $\mathscr{R}$(Mutating) = Organisms. *Example 2*. Define a binary relation $<$ on some set A. Then the reference class of $<$ is A. *Example 3*. Let $F: A \to B$ be a function. Since F is a particular case of a relation on $A \times B$, $\mathscr{R}_p(F) = A \cup B$.

DEFINITION 2.9 Let $\mathbb{P}$ be a family of n-ary predicates with domain $A_1 \times A_2 \times \cdots \times A_n$ and let S be the totality of statements formed with them. The function

$$\mathscr{R}_s : S \to \mathscr{P}(\bigcup_{1 \leqslant i \leqslant n} A_i)$$

is called the *statement reference function* iff it is defined for every s in S and satisfies the following conditions:

(i) The referents of an *atomic statement* are the arguments of the predicate concerned. More precisely, for every atomic formula $Pa_1 a_2 \ldots a_n$ in S

$$\mathscr{R}_s(Pa_1 a_2 \ldots a_n) = \{a_1, a_2, \ldots, a_n\}.$$

(ii) The reference class of an arbitrary *propositional compound* equals the union of the reference classes of its components. More exactly, if $s_1, s_2, \ldots, s_m$ are statements in S and if ω is an m-ary propositional operation,

$$\mathscr{R}_s(\omega(s_1, s_2, \ldots, s_m)) = \bigcup_{1 \leqslant j \leqslant m} \mathscr{R}_s(s_j).$$

(iii) The reference class of a *quantified formula* equals the reference class of the predicate occurring in the formula. More explicitly, if P is an n-ary predicate in $\mathbb{P}$, and the Q_i, for $1 \leqslant i \leqslant n$, are arbitrary quantifiers, then

$$\mathscr{R}_s((Q_1 x_1)(Q_2 x_2) \ldots (Q_n x_n) Px_1 x_2 \ldots x_n) = \mathscr{R}_p(P).$$

Example 1. $\mathscr{R}_s$(Venus is larger than Mars) = {Venus, Mars}.
Example 2. $\mathscr{R}_s(p \,\&\, q) = \mathscr{R}_s(p \vee q) = \mathscr{R}_s(p) \cup \mathscr{R}_s(q)$.
Example 3. $\mathscr{R}_s$(All humans are mammals) = Animal Kingdom.
More on this in the next subsection and in Sec. 5.2.

The relations among the various sets and functions involved in the preceding definitions are summarized in the diagram on the next page.

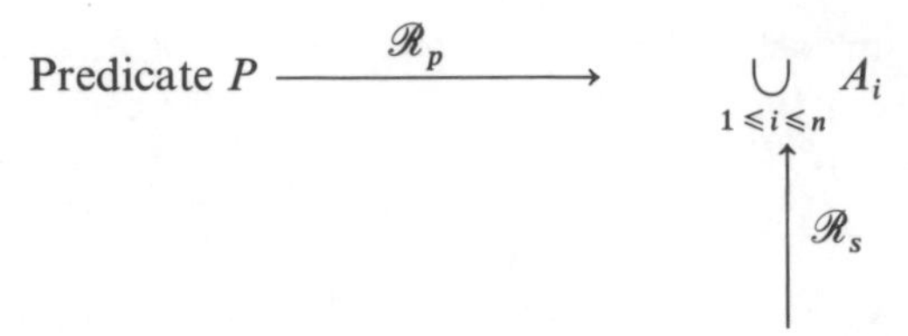

Remark 1 Constructs other than predicates and statements, such as the individual x and the set y occurring in the statement ⌜$x \in y$⌝, are assigned no referents in our theory. The reason is that individuals and collections thereof are to function as referents themselves. *Remark 2* According to Definition 9i, ⌜Pa⌝ is taken to concern the object named a, not the name 'a'. One might be tempted to introduce a new symbol, say 'a', for the nominatum of a. But this would start an infinite regress. *Remark 3* Our analysis of reference is applicable in all contexts, whether or not "extensional" (truth functional). For example,

$$\mathscr{R}(\text{Smith knows that } p) = \{\text{Smith}, p\}.$$

A binary epistemic predicate E, such as "knowing", "believing", or "doubting", will be analyzed as a function

$$E: \text{People} \times \text{Statements} \rightarrow \text{Statements},$$

whence

$$\mathscr{R}(E) = \text{People} \cup \text{Statements}.$$

There is nothing opaque about "oblique" reference if analyzed as multiple rather than single. More on "intensional" contexts in Ch. 4, Sec. 1.3. *Remark 4* Our definitions of the reference functions are necessary but not sufficient for a correct identification of the referents of a construct. A precise identification requires some fund of knowledge and it depends upon one's metaphysics. Consider, indeed, the statement that the weather is fine. The *ostensive* (or manifest or superficial) referent of this statement is the weather. But the weather is not a thing: it is a state of a thing. The *deep* (or hidden or genuine) referents of the given statement are the terrestrial atmosphere and mankind. In fact the given proposition may be construed as short for ⌜The atmosphere is in a state that suits humans⌝. The metaphysical principle behind this analysis of the given proposition is that the referents of factual constructs are concrete systems (things). For a radically different view concerning the relations of semantics to metaphysics, see Shwayder (1961).

3.3. *Some Consequences*

We proceed to drawing a few consequences from our Definitions 8 and 9 of the reference functions for predicates and for statements.

COROLLARY 2.1a The reference class of the equality (inequality) relation equals the set A on which $=$ is defined:

$$\mathscr{R}_p(=)=\mathscr{R}_p(\neq)=A.$$

COROLLARY 2.1b The reference class of a statement of inequality (equality) is constituted by the individuals concerned. That is, if $a, b \in A$, then

$$\mathscr{R}_s(a\neq b)=\{a, b\}, \qquad \mathscr{R}_s(a=b)=\{a\}.$$

COROLLARY 2.2a The reference class of the membership relation $\in$ equals the universe U of sets on which it is defined:

$$\mathscr{R}_p(\in)=U.$$

COROLLARY 2.2b The reference class of a statement of class membership is constituted by the relata concerned:

$$R_s(a\in A)=\{a, A\}$$

Remark. This last corollary gives rise to an apparent paradox. A membership statement is equivalent to a predicative statement:

$$Pa \quad \text{iff} \quad a\in\mathscr{E}(P), \quad \text{where} \quad \mathscr{E}(P)=\{x \mid Px\}$$

is the extension of the predicate. But $\mathscr{R}_s(Pa)=\{a\}\neq\mathscr{R}_s\,(a\in A)$. That is, *equivalents may have different reference classes*. And why should they unless equivalence be mistaken for identity or reference class for extension? ⌜Pa⌝ is equivalent to but not identical with ⌜$a\in\mathscr{E}(P)$⌝, if only because the predicates involved in each statement are not only of different rank and different type: while P applies to individuals, $\in$ relates individuals to sets and thus mixes objects of different types. Moral: The reference class we assign to a statement depends on the way we analyze the latter.

COROLLARY 2.3a The reference class of a set relation equals the union of the sets involved: If ρ is a binary relation between the sets A and B, then

$$\mathscr{R}_p(\rho)=A\cup B.$$

COROLLARY 2.3b The reference class of a statement of class relation consists of the classes involved:

$$\mathscr{R}_s(\rho AB)=\{A,B\}.$$

In particular

$$\mathscr{R}_s(A\subseteq B)=\{A, B\}$$

COROLLARY 2.4a The reference class of the conditional predicate $P\Rightarrow Q=\neg P\vee Q$ equals the domain shared by P and Q. More explicitly: if $P{:}A\rightarrow S$ and $Q{:}B\rightarrow S$ are unary predicates in $\mathbb{P}$, then because $P\Rightarrow Q{:}A\cap B\rightarrow S$,

$$\mathscr{R}_p(P\Rightarrow Q)=\mathscr{R}_p(P)\cap\mathscr{R}_p(Q)=A\cap B.$$

COROLLARY 2.4b The reference class of a universal conditional equals the intersection of the reference classes of the predicates involved:

$$\mathscr{R}_s((x)\,(Px\Rightarrow Qx)=\mathscr{R}_p(P)\cap\mathscr{R}_p(Q)).$$

Example The reference class of ⌜All ravens are black⌝, as well as of its negate, is the class of birds not that of black things nor any other one. Indeed, the predicate "$R\Rightarrow B$" is defined on the intersection of the domains of R and B, i.e. on $\mathscr{R}_p(R)$, because the domain of R is Birds and that of B is Things. More in Sec. 5.2.

COROLLARY 2.5a The reference class of a tautological predicate equals the union of the reference classes of the component predicates, whence it is not generally empty. In particular

(i) $\mathscr{R}_p(P\vee\neg P)=\mathscr{R}_p(P)$,
(ii) $\mathscr{R}_s((P)_{P\in\mathbb{P}}\,(P\vee\neg P))=\bigcup_{P\in\mathbb{P}}\mathscr{R}_p(P)$.

COROLLARY 2.5b The reference class of a tautological statement equals the union of the reference classes of the predicates involved, whence it is not generally empty. In particular

(i) $\mathscr{R}_s((x)(Px \vee \neg Px)) = \mathscr{R}_p(P)$,
(ii) $\mathscr{R}_s((P)_{P \in \mathbb{P}}(x)(Px \vee \neg Px) = \bigcup_{P \in \mathbb{P}} \mathscr{R}_p(P)$.

Remark 1 The tautology in (i) is *valid in the domain* of P, while the one in (ii) is valid in the total domain of the predicate family $\mathbb{P}$, i.e. *valid* without further ado. *Remark 2* Reference class and extension coincide only in the case of tautological predicates, or of tautologies, involving unary predicates. This coincidence seems to be an important source of the common confusion between reference class and extension.

COROLLARY 2.6a The reference class of a predicate and its negate are the same: If P is in $\mathbb{P}$, then

$$\mathscr{R}_p(P) = \mathscr{R}_p(\neg P).$$

COROLLARY 2.6b The reference class of a statement and of its denial are the same: if p is in S, then

$$\mathscr{R}_s(p) = \mathscr{R}_s(\neg p).$$

Example $\mathscr{R}_s$ (Some people are cruel) $= \mathscr{R}_s$ (Nobody is cruel) $=$ People.

COROLLARY 2.7 The reference class of a singular tautological statement, or of its negate (the corresponding contradiction), equals the union of the reference classes of its component. In particular

$$\mathscr{R}_s(p \vee \neg p) = \mathscr{R}_s(p \,\&\, \neg p) = \mathscr{R}_s(p) = \mathscr{R}_s(\neg p).$$

Remark 1 It is often claimed that analytic statements do not inform about the world because they are not about it. While the claim is right the reason given in support of it is dubious. ⌜Antarctica is cold or it is not cold⌝ concerns Antarctica, both for common sense and on our semantics, even though it "says" nothing, i.e. its factual content is nil. It is for the latter reason, not because of its alleged lack of reference, that the tautology is not affected by the recent finding that Antarctica was not cold during the Triassic. The same holds for analytically false statements: if singular they refer even if they do not inform. *Remark 2* Likewise for tautological predicates: since they hold for anything they refer to everything. *Remark 3* The preceding consideration would seem to contradict the thesis of the autonomy of logic espoused in Sec. 2.3. It does not, for what is peculiar to tautological singular statements is their *validity irrespective of their reference*, hence the impossibility of checking them against experimental evidence. *Remark 4* Since any given tautology is equivalent to every other tautology, it would seem that Corollary 7 leads to contradiction. It does not because we have refrained from requiring that equivalents be coreferential. Recall the comment to Corollary 2b. *Remark 5* The preceding remarks show the need for a concept of strong equivalence, or equivalence both syntactical and semantical. A possible definition is this: Two statements are said to be *strongly equivalent* iff they are equivalent and have the same referents. However, we shall not pursue this matter here. *Remark 6* Our reference functions refer to (are defined on) predicates and statements not on signs such as predicate letters and sentences. While on Definitions 8 and 9 every predicate or statement is referential (even if it concerns the null individual), not every expression denotes. A nonsense sign such as '&(%+' presumably occurs in no conceptual language, hence it designates nothing, whence it denotes nothing. (But of course it may suggest something – e.g. that the linotype ran amok – and may thus have a pragmatic meaning.) Even sophisticated signs composed of individually significant marks in a way that looks well formed may fail to designate constructs. (Two famous frauds are '1/0' and '$\{x \mid \neg(x \in x)\}$'. These are not, as is sometimes held, irreferential concepts: they are just nonconceptual signs.) *Remark 7* Even logical predicates refer – namely to constructs. For example, the reference class of the conjunction is the set of all the statements on which it is defined. This follows from Definition 8 applied to & construed as a certain function $\&: S \times S \rightarrow S$.

What follows will not depend critically on the precise formulation of our definitions of $\mathscr{R}_p$ and $\mathscr{R}_s$: in most cases it will suffice to assume that these functions are somehow well defined. Moreover, in the sequel we shall hardly distinguish between these two functions, for we shall be concerned with constructs in general.

3.4. *Context and Coreference*

The concepts of reference functions and reference class allow us to elucidate a number of other concepts. For example the following notions.

DEFINITION 2.10 The ordered triple $\mathbb{C}=\langle S, \mathbb{P}, D\rangle$ is called a (conceptual) *context* (or *frame*) iff S is a set of statements in which only the predicate constants in the predicate family $\mathbb{P}$ occur, and the reference class of every P in $\mathbb{P}$ is included in the universe or domain $D \subseteq \Omega$.

DEFINITION 2.11 c is called an *individual construct* in the context of frame $\mathbb{C}=\langle S, \mathbb{P}, D\rangle$ iff (i) c is either in $\mathbb{P}$ or in S, and (ii) the reference class of c is a unit set or singleton.

DEFINITION 2.12 c is called a *universal construct* in the context or frame $\mathbb{C}=\langle S, \mathbb{P}, D\rangle$ iff c is in $\mathbb{P}$ or in S but is not an individual construct.

Remark The preceding definitions of individual and universal relativize these concepts to a definite conceptual context or frame. This is as it should be, for one and the same object may be regarded as an individual in a given context and as a set in another. And a property may concern every individual in a class but not all individuals. This gives some precision to the classical distinction formulated by the Philosopher: "By the term 'universal' I mean that which is of such a nature as to be predicated of many subjects, by 'individual' that which is not thus predicated. Thus 'man' is a universal, 'Callias' an individual" (*On Interpretation*, Ch. 7).

DEFINITION 2.13 c is said to be an *individual construct* iff c is an individual construct in every context.

DEFINITION 2.14 c is said to be a *universal construct* iff c is a universal construct in every context.

Whether these last two concepts have any application must be left to other disciplines to find out. (Perhaps their only use is to occur in the conjecture that there are no such constructs.) Let us now turn to the comparison of reference classes.

DEFINITION 2.15 Two constructs are called *coreferential* (or *equirefetial*) in a context $\mathbb{C}=\langle S, \mathbb{P}, D\rangle$ iff (i) they belong to the context $\mathbb{C}$ and (ii) they have the same reference class. Symbolically:

$$\text{If } c \text{ and } c' \text{ are in } \mathbb{C}, \text{ then } c \sim_r c' =_{df} \mathcal{R}(c)=\mathcal{R}(c').$$

Remark The relativization to a context is not really needed if one proceeds carefully – i.e. with hindsight. For example, in the context of nonrelativistic physics the concepts of mass and charge are equireferential: both concern only bodies. But in relativistic physics they cease to be coreferential (although they are still referentially commensurable). For, although the two concepts concern bodies, that of mass concerns now also reference frames, and it is possible to make certain fields serve as frames. Of course two different concepts of mass are involved in the two contexts even though the two are usually designated by the same word. Should this be overlooked, the relativization to a context would avoid further confusion. (More in Sec. 4.2.)

As is clear from Definition 15, the relation $\sim_r$ is an equivalence relation. As such it enables us to form equivalence classes. Thus it allows us to form

DEFINITION 2.16 The set $[c]_{\mathbb{C}}$ of *coreferentials* of a given construct c in a context $\mathbb{C}$ is

$$[c]_{\mathbb{C}} =_{df} \{c' \in \mathbb{C} \mid c' \sim_r c\}.$$

Example All of the theories of solids, no matter how different, have the same reference class, namely the set of solid bodies.

Being an equivalence relation, equireference induces a partition of the totality C of constructs, i.e. an exhaustive division of C into mutually disjoint classes of coreferentials. This partition or collection of equivalence classes is the quotient set of C by $\sim_r$, or $C/\sim_r$ for short. We may call $\Pi = C/\sim_r$ the *referential partition* of the conceptual universe. This concept allows us to elucidate the intuitive concept of semantic homogeneity of a theory (Bunge, 1967a, Ch. 7) by means of

DEFINITION 2.17 A theory will be said to be *referentially homogeneous* iff it is included in an element of the referential partition $C/\sim_r$, i.e. if its constructs are all in one of the equivalence classes of equireferentials.

DEFINITION 2.18 A theory will be said to be *referentially heterogeneous* iff it is not referentially homogeneous.

Example Psychology and physics are referentially heterogeneous. On the other hand, according to the identity theory, psychology and

neurology are referentially homogeneous since both concern animals.
Let us next specialize the reference functions.

4. Factual reference

4.1. *The Factual Reference Class*

In the so-called empirical sciences we are particularly interested in a special concept of reference, namely reference to concrete (though perhaps hypothetical) existents. We therefore introduce

DEFINITION 2.19 Let $\mathscr{R}(c)=A_1 \cup A_2 \cup \ldots \cup A_f \ldots \cup A_n$ be the reference class of a construct c, where the A_i for i between 1 and $f<n$ are sets of non-conceptual items, i.e. such that $A_i \nsubseteq C$. Then the *factual reference class* of c is the union of the classes of factual items:

$$\mathscr{R}_F(c)=A_1 \cup A_2 \cup \ldots \cup A_f \subseteq \mathscr{R}(c).$$

Example 1 $\mathscr{R}_F$ (Light rays are represented by straight lines) = Light rays. The total reference class includes also the set of straight lines. *Example 2* Let $f: A \to B$ be a function from a set A of concrete systems to a set B of numbers. For example, f could represent the relative dilation of bodies. Then $\mathscr{R}_F(f)=A$, whereas $\mathscr{R}_p(f)=A \cup B$. *Example 3* Let $g: A \times B \to R$, where A is a set of concrete systems, B the set of conceivable scale-cum-unit systems associated to the magnitude g, and R the set of real numbers. Since B is conventional, it is included in the class of constructs, so that we are left with $\mathscr{R}_F(g)=A$.

Not all the constructs occurring in factual science have a factual reference. For one thing the logical concepts, such as "not" and "all", have no such reference. Nor do scale concepts, proportionality constants, and dimensional constants have referents (recall Sec. 2.4). It will therefore be convenient to coin a name for such constructs. The following conventions will do.

DEFINITION 2.20 A construct which is either tautologous or has an empty factual reference class is said to be *factually vacuous*.

DEFINITION 2.21 A construct which is not factually vacuous, i.e. which is neither tautologous nor has a void factual reference class, is called *factual*.

Example 1 ⌜1 m = 100 cm⌝ is factually vacuous. *Example 2* The state function ψ is a factual construct because it concerns physical entities. *Example 3* The semantic assumption ⌜ψ represents the state of an individual microsystem⌝ is a factual construct because it is nontautologous and includes the factual concept ψ.

Let us now restrict the concept of a context introduced by Definition 10:

DEFINITION 2.22 The triple $\mathbb{C}_F = \langle S, \mathbb{P}, D \rangle$ is called a *factual context* iff (i) it is a context and (ii) $\mathbb{P}$ includes a nonempty subset of factual predicates.

DEFINITION 2.23 Any context that fails to be factual is called a *formal context.*

DEFINITION 2.24 Let $\mathbb{C}_F = \langle S, \mathbb{P}, D \rangle$ be a factual context. The predicate in $\mathbb{P}$ with the largest factual reference class is called *maximal* in $\mathbb{C}_F$. Symbol: $P_{\max}$.

Remark 1 Logic, mathematics and semantics are formal contexts since they deal with constructs. On the other hand history, linguistics, epistemology and metaphysics are factual contexts. Mythology, religion and science fiction are factual contexts too. Their extension or domain of truth is another matter. *Remark 2* Definition 23 effects a dichotomy in the set of contexts. However, the factual/formal dichotomy is not universally accepted: to some even mathematics is *au fond* factual or empirical, while a few hold science to be mathematics with a narrow scope. The semantic method for settling this issue in particular cases is to perform a semantic analysis of the basic concepts of the context in dispute. If at least one such concept has a factual (nonconceptual) reference class then the context is factual – otherwise formal. This semantic analysis can be checked and supplemented by a methodological analysis: if at least one validation procedure is empirical (e.g. experimental) then the context is factual. But of course for such an empirical test to be possible the context must be factual to begin with. (Caution: computer simulations do not qualify as empirical even if they bear on factual contexts.) *Remark 3* We saw in Sec. 3.3 that singular tautologies, like ⌜Antarctica is cold or not cold⌝, can have a factual referent. It would seem then that logic is a factual context after all, contrary to our previous contention. Not so, because logic is concerned with the whole of the conceptual universe C, in particular with statements – not with their referents. (E.g. the reference class of a propositional connective is the class of statements on which it is defined.) And, as we saw in Sec. 2.1, the reference relation is not generally transitive.

Of all the factual contexts the most interesting ones, because the richest, are the factual sciences. A *factual science* is a science with a factual reference, i.e. it is a factual context capable of being tested in accordance with the scientific method. This quick characterization of a factual science is of a methodological character. A strictly semantic definition of science seems to be out of the question: a precise sense and a factual reference are necessary but insufficient to have a science, and a high degree of truth, though desirable, is neither necessary nor sufficient. For this reason we do not give here a formal definition of factual science: this is a matter for methodology (cf. Bunge, 1967a). But once we have settled for some characterization of factual science we can use our semantics to shed some light on certain aspects of the former. First of all the concept of subject matter of a scientific subject:

DEFINITION 2.25 Let $\mathscr{S}_F = \langle S, \mathbb{P}, D \rangle$ be a factual science. Then the *domain* (or *subject matter*) of $\mathscr{S}_F$ equals the factual reference class of the maximal predicate $P_{\max}$ in $\mathbb{P}$:

$$Domain\ (\mathscr{S}_F) = \mathscr{R}_F\ (P_{\max}).$$

Needless to say, a science need not have a homogeneous domain in the sense of Definition 3 in Sec. 2.3. And a science need have no precise deductive structure: it can be just a context, which is something slightly more structured than a mere set of statements, in that a given family of predicates is shared.

And now to the central referent of our semantics:

DEFINITION 2.26 T is a *factual theory* iff (i) T is a theory and (ii) T contains factual predicates.

We leave metamathematics, in particular the theory of theories, the elucidation of the concept of a theory. Our present concern is with the objects a factual theory is about:

DEFINITION 2.27 The *factual reference class*, or *universe of discourse*, of a theory T equals the reference class of the maximal factual predicate P_{max} in T:

$$Domain\ (T) = \mathscr{R}_F\ (P_{max})$$

Remark 1 Since theories in logic and pure mathematics contain no factual predicates, their factual reference classes are empty. *Remark 2* The factual reference class of a scientific theory is constituted by possibles not just actuals. For example, the genetics of a given biological species is about all possible genotypes of that species – only a tiny fraction of which are ever actualized. *Remark 3* The notion of a possible individual, or a possible factual item, is sometimes regarded as ghostly and at other times in need of modal logic. Yet it is not ghostly, for science handles it every day: thus mechanics is the study of the possible motions of possible bodies. And the concept cannot be characterized by modal logic, which is too poor a theory. But is easily defined in our semantics, namely thus: The factual item x is *possible according to the theory* T iff x belongs to the factual reference class *Domain* (T) of T. (In science we have no use for the absolute modalities handled by modal logics.)

4.2. *The Factual Reference Class of Scientific Theories*

The determination of the reference class of a factual theory is far from being a straightforward business. Scientific theories come usually unkempt: they are more or less haphazard collections of formulas accompanied by extrasystematic remarks. These hints are ambiguous and are often loaded with an obsolete philosophy, namely, operationism, which forces the observer, or experimenter, into every reference class. Our problem has a solution only if the theory is axiomatized with respect to both its form and its reference. We shall henceforth assume that the scientific theories which are the object of our analysis have been axiomatized in both respects. Not that this is necessary in order to use the concept of factual reference of a theory, but it is needed for computing the reference class with precision.

If the theory is formal, its axioms will determine the structure of the basic (primitive or undefined) concepts as well as their interrelations. For example, if one of the basic concepts of the theory is a binary associative operation on a certain set, then one of the axioms of the theory will state this much. But if the theory is factual such a formal characterization will be insufficient: it must be supplemented with semantic assumptions indicating the nature of the referents – whether they are seeds, or birds, or perhaps both. In other words, a factual theory will be a formal structure enriched with a set of assumptions concerning the referents of its basic concepts. In a nutshell, we stipulate

DEFINITION 2.28 T is an *axiomatic factual theory* iff (i) T is a factual theory (according to Definition 26), (ii) T is axiomatized and (iii) every basic (undefined) factual concept of T is explicitly assigned a definite nonempty reference class by some of the axioms of T.

This is the kind of axiomatics Hilbert and Bernays (1968) called *inhaltlich*, as opposed to *formal* axiomatics, which abstracts from the specific content of the original theory. An axiomatic factual theory is determined by both its formal (mathematical) axioms and by the nonformal (semantical) assumptions of the theory – roughly those that indicate the referents of the basic predicates in the theory. The first part of the preceding statement is a truism from the theory of deductive systems (Tarski, 1956, Ch. xii). The second part will be proved. As to the first part: calling $\mathscr{A}$ the axiom base of T, we symbolize that statement in the standard way

$$T = \mathscr{C}n(\mathscr{A})$$

which can be read: 'an axiomatic theory is the totality of consequences of its axioms'. In addition, the (nonformal) axioms of a factual theory determine not just the referents of its factual predicates but the reference class of the theory as a whole (as characterized by Definition 27). To show this we must start by construing the concept of an axiom base as the conjunction of all the axioms of the theory. We shall limit ourselves to a finitely axiomatizable theory, but infinitary logic can handle the infinite case. We stipulate then

DEFINITION 2.29 Let A_i, with $1 \leqslant i \leqslant n$, be the n axioms of a theory T. Then the *axiom base* of T is

$$\mathscr{A}(T) =_{df} \bigwedge_{i=1}^{n} A_i.$$

We can now state

THEOREM 2.1 Let $\mathscr{A}$ be the axiom base of a theory T. Then

$$\mathscr{R}(T)=\mathscr{R}(\mathscr{C}n(\mathscr{A})).$$

Proof sketch In an axiomatic theory no theorem contains predicates other than those occurring in its axiom base. In other words, every concept in an axiomatic theory is either defining (primitive) or defined. Hence deduction cannot change the kind or number of referents assigned the basic factual predicates.

THEOREM 2.2 The factual reference class (or universe of discourse) of an axiomatic theory equals the union of the factual reference classes of its axioms:

If $\mathscr{A}(T)=\bigwedge_{i=1}^{n} A_i$, then $\mathscr{R}_F(T)=\bigcup_{i=1}^{n} \mathscr{R}_F(A_i)$.

Proof By Theorem 1, Definition 9, and Definition 19.

Remark 1 The last theorem is the most important in our theory of reference. It may be called the theorem of *reference conservation* or reference invariance under deduction. It constitutes a partial justification of the dictum that the conclusions must be "contained" in the premises – a dictum that, as we shall presently see, is not generally valid. *Remark 2* The preceding theorem is far from trite: in fact it fails for nonaxiomatic theories. Indeed, in the latter new predicates, totally alien to the basic ones, can be adjoined at any step. This adjunction can be made in either of two ways: via the introduction of new assumptions or via deduction with the help of the principle of addition of formal logic. The former move is obvious: informal theories are forever under construction or repair. The second move is less obvious and deserves special notice. *Remark 3* Let t be a formula of an informal theory T. By the principle of addition, $t \vdash t \vee u$, where u is an arbitrary statement, i.e. one that need bear no semantical relation to t. In fact, as far as logic is concerned, the reference classes of t and u need not overlap, let alone coincide – i.e. t and u may fail to be coreferential. Thus t could be about stars and u about frogs – and yet u could be counted among the members of the open theory T. In general: in an open context, the reference of a consequence may be wider than that of the premises. This violates the principle of the Aristotelian organon according to which "the indemonstrable premises of a science (...) must be within one genus with its conclusions" (*Posterior Analytics*, Bk. I, Ch. 28, 88). If this principle, tacitly obeyed in most cases, is surrendered then anything can be claimed to be relevant to anything else. The result will be confusion or error or both. Such a move, hardly avoidable in informal theories (which are open contexts), is automatically precluded in axiomatic theories, which are closed contexts. While the former are semantically loose or open, hence both perfectible and corruptible, axiomatic theories are semantically tight or closed. This last expression is elucidated by

DEFINITION 2.30 A set S of statements will be said to be *semantically tight*, or *closed*, iff S has the same referents as its logical consequences, i.e. if

$$\mathscr{R}(S)=\mathscr{R}(\mathscr{C}n(S)).$$

Theorem 1 can now be restated in this way: *Axiom systems are semantically closed.* For this reason the axiomatic format, though usually unsuitable during the early construction stage, is mandatory for investigating the semantics of a theory: *it is only in axiomatic contexts that we know exactly what we are talking about.* This is because axiomatization involves fixing in advance the set of basic predicates, hence the universe of discourse, to be admitted in the given context. Thus axiomatization suffices to get conclusions that are semantically *relevant* to the premises. Should it be objected that the price to pay for semantic coherence or relevance – namely axiomatization – is too high, it may be rejoined as follows. First, price is irrelevant. Second, axiomatization is a means for attaining a number of other goals as well (see Bunge, 1973b). Third, the alternative procedure for trying to secure relevance is far more expensive than axiomatization, as it consists in adopting some system of relevant implication (cf. Anderson, 1972). This change in logic involves not only new rules of inference but also modalities – none of which seem to be called for by scientific work. And, despite commanding such a high price, this alternative procedure does not really guarantee semantic relevance, as it offers no criteria for it and it places no restrictions upon the principle of addition.

If we wish deduction to preserve reference even outside semantic contexts then we must keep a permanent watch on the consequences and screen out all those that violate semantic closure, i.e. that are not referentially germane to the premises. (One of the rare occasions on which such a precaution is taken explicitly, is in stating Craig's interpolation theorem. A standard formulation of this theorem is as follows. Let p and q be formulas such that $\vdash p \Rightarrow q$. Then there exists a third formula r, *containing only predicates occurring in p and q*, such that $\vdash p \Rightarrow r$ and $\vdash r \Rightarrow q$.) The very choice of premises in an argument should be regulated by the unspoken yet time honored *principle of relevance*: The premises should be not only mutually compatible in the formal sense but also semantically congenial. This principle of argumentation limits the range of applicability of the so called rule of augmentation of premises, according to which if p entails q then q follows also from p conjoined with an arbitrary premise r. This rule holds only if the additional premise is either idle or both formally and semantically coherent with the original premises. Once the initial assumptions have been chosen we may use a deduction technique that complies automatically with the requisite of semantical closure and has the additional advantage that it does not require guessing the conclusions beforehand, so that it may be used by a nonmathematician – e.g. a computer. (See, e.g., Hilbert and Ackermann, 1950, p. 24.) This *mechanical procedure for extracting reference preserving conclusions* is as follows. (i) Conjoin all the assumptions; (ii) expand the conjunction in conjunctive normal form, i.e. as a conjunction of binary disjunctions; (iii) detach every conjunct and every conjunction of such. The detached formulas constitute the desired set, i.e. the maximal set of consequences compatible with reference conservation. This subset of consequences is semantically closed and finite, hence a smallish subset of the infinite set of all the consequences of the given premises. We cannot have both all the infinitely many consequences of a finite set of premises, and semantical closure.

Our Theorem 2 of reference conservation allows us to determine the reference class of any factual theory cast in an axiomatic format and consequently it enables us to compare theories as to their domains. In fact the theorem backs up the following conventions.

DEFINITION 2.31 Let T and T' be two axiomatic factual theories. Then T and T' are said to be *referentially commensurable* iff their reference classes overlap, i.e. if

$$\mathscr{R}_F(T) \cap \mathscr{R}_F(T') \neq \emptyset.$$

DEFINITION 2.32 Let T and T' be two axiomatic factual theories. Then T' is said to be *referentially bulkier* than T iff

$$\mathscr{R}_F(T') \supseteq \mathscr{R}_F(T).$$

Example 1 Particle mechanics and the theory of electromagnetic fields in a vacuum are incommensurable: they are not about the same things. *Example 2* Electrodynamics and mechanics are referentially commensurable: they have common referents – bodies. *Example 3* Relativistic mechanics is referentially bulkier than nonrelativistic mechanics because, unlike the latter, it concerns both bodies and electromagnetic fields. Hence they are referentially commensurable. Consequently they are comparable – *pace* Kuhn and Feyerabend.

Remark 1 If two theories are referentially incomparable then they are not comparable at all except in strictly formal respects. In particular they cannot be compared as to explanatory and predictive powers: they do not explain and predict any shared facts. *Remark 2* The concept of referential commensurability will be supplemented with that of methodological commensurability in Sec. 5.1. *Remark 3* Our concept of referential commensurability is at variance with Kuhn's (1962). For Kuhn two theories are incomparable iff they fit different theory paradigms and thus are conceptually different even though they may concern the same things. Behaviorist and cognitivist learning theories would be incommensurable in this sense. But, since they concern the same kinds of animal, to us they are referentially commensurable and thus comparable. If they were not they could not be regarded as rival. *Remark 4* Nor does our criterion of theory commensurability agree with Feyerabend's, which is the sharing of *statements*, hence of concepts. It is of course true that relativistic mechanics "does not, and cannot, share a single statement with its predecessor" (Feyerabend, 1970, p. 82). But it is also true that both refer to bodies (cf. Example 3 above). Moreover the basic concepts of classical mechanics can be retrieved from the corresponding relativistic concepts: for example the classical mass function is a certain restriction of the relativistic mass function (see Bunge, 1970b). Therefore the two theories are commensurable, hence comparable – which is why they are being compared all the time. If this were not the case then there would be no more ground for preferring one of the theories over the other than for preferring Newton to Vivaldi. *Remark 5* The overlapping of two theories, i.e., their sharing some statements (Feyerabend's criterion) is not only unnecessary for their commensurability: it is not even sufficient. In fact two factual theories with different referents may have the same mathematical formalism, as is the case with certain theories of epidemics and of rumor propagation. It would be preposterous to declare such theories semantically commensurable even though they share infinitely many statements. What is decisive (necessary and sufficient) for theory commensurability is not the mere sharing of statements but the sharing of semantic assumptions, for these are the ones that point to the referents or their properties. *Remark 6* The moral of the preceding remarks is this. The interesting problem of comparing rival scientific theories and ascertaining their meaning diversity cannot be approached on the basis of spotty historical remarks: it requires a full fledged semantics of science and an analysis of scientific theories in the light of the former.

4.3. *Spotting the Factual Referents: Genuine and Spurious*

The preceding examination of the factual reference class of a scientific theory presupposes that the latter has been formulated in a correct way, both formally and semantically, so that finding its referents is a straightforward operation. In other words, we have assumed that the theory to be analyzed contains detailed analyses of every one of its variables: in this case Theorem 2 solves our problem. Unfortunately this situation is exceptional rather than typical: in most cases the referents of a theory are hinted at in extrasystematic remarks dropped in motivating the theory or, as we saw in Sec. 1, they are the subject of spirited controversy.

In other words, semantics can be of little help in spotting the genuine referents of a theory unless the latter has already been axiomatized in such a way that those referents are clearly exhibited. But at least semantics can help us to dig up the referents of an untidy theory and can assist us in formulating the semantic assumptions of the theory. These tasks are performed in the light of a criterion that, unlike Theorem 2, concerns unkempt theories. Here it is:

CRITERION An object is a *genuine factual referent* of a scientific theory iff it is involved in at least one of the law statements in the theory.

Otherwise, i.e. if an object falls under no law of the theory, then it is a *spurious referent* of it, no matter how many authorities may claim the opposite. For example, if a theory of electrons fails to contain law statements about instruments sensitive to electrons, then the theory does not refer to such instruments, much less to those in charge of the latter. It will not do to claim on philosophical grounds that some reference to instruments or to observers *must* be "implied" or "presupposed" by the theory, for otherwise the latter would be neither meaningful nor testable.

The preceding criterion suggests adopting the following

DEFINITION 2.33 A nonconceptual object x is a *genuine factual referent* of an informal theory T iff (i) T contains at least one law statement involving one or more variables concerning x; (ii) such law statements cease to hold upon the removal of the variable(s) in question, and (iii) T can be adjoined, without introducing any contradiction, assumptions effecting a more precise characterization of the variable(s) referring to x.

Whatever passes for referent but is not genuine is a *spurious referent*. Contemporary science contains several theories with spurious referents, i.e. that purport to but in fact do not refer to certain entities. Such theories may be called *ghostly*. We proceed to exhibit three ghosts.

Example 1 Probability is sometimes interpreted in a subjectivist or semisubjectivist way, occasionally even in the context of natural science. For instance, it is sometimes held (e.g. by Brillouin 1962) that the distribution function (or else the partition function) in statistical mechanics, and the various probability densities in quantum mechanics, concern our own information (or rather want of it) instead of either objective (though potential) properties or objective random distributions. If this contention were true then statistical mechanics and quantum mechanics would concern the cognitive subject and his mental states at least as much as autonomous physical entities. By our Criterion, if the cognitive subject were a genuine referent of those theories then these should contain law statements enabling us to explain and predict our own behavior. But they do not. Consequently the cognitive subject is a spurious referent of those theories. Equivalently: those theories are about autonomous physical systems. This conclusion is reinforced by the consideration that one learns about such systems by studying them rather than by the method of transcendental meditation. *Example 2* The so called measurement theories found in the foundations of mathematical psychology happen to be concerned with certain types of scientific concept not with the empirical operation scientists call 'measurement'. So much so that they do not involve any laws of either the measured object or the measuring instrument, whence they are of no help in designing and interpreting any measurement proper or even in estimating random errors of measurement (see Bunge, 1973c). *Example 3* It is usually held that the quantum theory of measurement (or some variant of it) accounts for actual measurements and moreover concerns the system-apparatus-observer compound. However, this theory fails to contain variables representing any properties of the observer. *A fortiori* the latter does not occur among the referents of the law statements of the theory – to the point that the author of the standard quantum theory of "measurement" acknowledged that the observer "remains outside the calculation" in his theory (von Neumann, 1932, pp. 224, 234). And although variables alleged to represent properties of an instrument do play a role in this theory, they fail to concern real instruments. As a consequence none of the formulas of this theory predicts any experimental results. A genuine measurement theory concerns a specific set-up, hence it contains specific law statements accounting for the peculiar structure and composition of the apparatus and its coupling with the measured thing. Such a specificity shows up in the constitutive equations or, equivalently, in the occurrence of non-universal parameters and constants, such as the refractive index, the electric resistivity, or the magnetic permeability. The quantum theory of "measurement" contains no such constitutive equations, hence it can refer to no genuine measurement set up. In sum, the quantum theory of "measurement" is ghostly: both the observer and the apparatus are spurious referents of it. (For details see Bunge, 1967b, Ch. 5; 1973b, Ch. 4.)

4.4. *The Strife over Realism in the Philosophy of Contemporary Physics*

The assumed referents of a scientific theory may be physical, chemical, biological, social, or what have you. Whatever their kind, the referents can be regarded in a number of alternative ways in their relation to the cognitive subject. The main doctrines that dominate the current philosophy of science in this respect can be summarized as follows.

(i) *Conventionalism* Scientific theories have no referents: they are just data summarizing and data processing tools – in particular they are prediction devices.

(ii) *Pragmatism, operationism, phenomenalism* Scientific theories are about phenomena, i.e. facts in which some cognitive subject is involved and such as they appear to him: they concern neither the subject alone (subjectivism) nor an autonomous object (realism).

(iii) *Realism* If the referents of a scientific theory exist at all then they exist on their own, i.e. independently of their being perceived or conceived: the cognitive subject creates, tests and applies hypotheses and theories instead of either creating their referents or posing as the latter.

The defense of any such stand is usually dogmatic: it consists in quoting authorities. At most it is pragmatic: it derives its strength from the heuristic power of the given philosophical doctrine, as when operationism is defended on the ground that it helps in the conceiving of experiments, or when realism is tolerated because it assists to invent "theoretical entities". Such lines of defense are weak and they foster sterile controversy. We need a different strategy, one relying on a semantical and methodological analysis of the theories concerned. We should be able to use a theory of reference to disclose the genuine referents of a scientific theory and thus evaluate the conflicting doctrines concerning such referents. How this can be done will be sketched presently.

From a semantic viewpoint conventionalism is false, since a theory with no factual reference whatever may qualify as a mathematical theory, hardly as a scientific theory. It is true that most of the formulas of a scientific theory do not involve explicitly the object variables or referents discussed in Sec. 2.4, so that the formulas look reference free, as if they were formulas in pure mathematics. But the factual reference is usually indicated or suggested by the accompanying text, sometimes called 'the prose of the theory'. And when these extrasystematic remarks are incorporated systematically into the theory, as so many semantic assumptions, the referents are brought to light. In short conventionalism, plausible as it may look in relation to the bare mathematical bones of scientific theories, is mistaken in relation to suitably axiomatized factual theories. So much for a semantic criticism of conventionalism.

The preceding semantic considerations are reinforced by the following methodological remarks. First, by refusing to make a semantic (or referential) commitment, conventionalism fails to draw the line between the various sciences and even between science and mathematics. (More on semantic formalism in Ch. 6, Sec. 3.1.) Second and consequently, conventionalism is incapable of telling us what kind of data are relevant to a given theory. Third and consequently, conventionalism does not help us to devise empirical tests, as every test of a scientific theory presupposes our knowing its reference class. Therefore the conventionalist view that scientific theories are data processing contraptions fails to do justice to both theories and data.

If conventionalism is discarded then only two serious contenders remain: phenomenalism and realism. (The truth might of course belong to a further party not involved in the current strife, but it will turn out that semantics and methodology favor just one of the current rivals.) Before the days of relativity and quanta, physics was usually interpreted in a realist way. Thus the length $L(b)$ of a bar b (or rather its length relative to a length standard) was regarded as an intrinsic and objective property of the body b. With the advent of special relativity (1905) length became a joint or mutual property or two physical systems: the body b and a reference frame f. Absolute, i.e. frame-free lengths, were swept aside in principle though not in practice. In other words, the definite description '$L(b)$' was replaced by '$L(b, f)$', short for "the length of b in (or relative to) the frame f". That is, a new length function was introduced: one with a different reference class. While the reference class of the nonrelativistic length concept was the set of physical systems, the reference class of the relativistic length concept is the union of the set of physical systems and the set of reference frames.

Now, the term 'reference frame' may be construed in either of two ways: as a physical system of a special kind, or as an observer. If the former, then nothing changes from an epistemological point of view: a property that had been regarded as absolute (or frame independent) turned out to be relative (or frame dependent). Equivalently: a relational metric property was substituted for an intrinsic one. But if reference frames are conceived as sentient beings equipped with yardsticks, clocks, mirrors, and other pieces of apparatus, then length ceases to be a strictly physical property to become a property of compound systems with a human component. And this is epistemologically revolutionary – or rather counter-revolutionary, as it reinstates anthropocentrism. Thus the expressions 'apparent length' and 'the length measured by a moving observer' were coined to indicate that the theory concerns appearances (to observers) not autonomous reality. The facts that all basic law statements are required to be frame free (or covariant), and that many magnitudes (such as the spacetime interval, the electric charge, and the entropy) are invariant, i.e. the same in (relative to) all frames, were not taken as evidence against this operationist interpretation. They were just ignored.

Two decades later quantum mechanics suffered a similar fate: its formulas were read in the light of the then dominant philosophy of physics. According to the usual or Copenhagen interpretation of quantum mechanics, every formula of the theory is about some microsystem under the action of an experimental set up controlled in an arbitrary fashion by an observer. Because of this alleged control and of the supposed docility of microsystems, which are imagined to behave as commanded by the observer, the apparatus-observer complex is usually called "the observer" – even if the whole experiment is automated. And dynamical variables are christened "observables". In this way no objective physical properties would seem to remain: they all become, just as with the operationist interpretation of relativity, observer dependent. For example, an eigenvalue is construed as a possible measurement result, an eigenstate as the corresponding observed state, a superposition of eigenstates as symbolizing our own uncertainty concerning the state of the system for not being under observation, and so forth. Consequently phenomenalism is pronounced victorious and realism dead. From here on the transition to strait subjectivism is easy: "the basic principles of physics, embodied in quantum-mechanical theory, deal with connections between observations, that is, contents of consciousness" (Wigner, 1970).

Do these non-realist interpretations of contemporary physics have any ground other than authority? Could semantics help us to make a rational choice between phenomenalism and realism? It can indeed, namely by supplying the criterion and the definition of factual reference given in Sec. 4.3. To reveal the genuine factual referents of a theory pick its basic predicates, analyze them and disclose the way they function in the central law statements of the theory – i.e. before the latter is applied or tested. Of all the alleged independent variables only those will qualify as genuine which are characterized by the theory and occur in the latter's law statements: all others make no difference and thus are ghostly, i.e. they concern spurious referents. Let us see how this criterion works in weeding out ghostly variables in a couple of important cases.

Example 1 Consider the famous Lorentz "contraction" formula

$$L(b, f) = L(b, b)\,[1 - u^2(b, f)/c^2(w)]^{1/2} \qquad \text{(SR)}$$

where b denotes a body (in fact any physical system), f a reference frame, and w an electromagnetic wave in a vacuum. The formula gives the length of b in (relative to) f, in terms of the length $L(b, b)$ of b relative to b itself, and of the ratio $u(b, f)/c(w)$ of two velocities: the velocity of the body with respect to the frame, and the (absolute) speed of light $c(w)$ in void. The phenomenalist construes f as an observer and interprets $L(b, f)$ as the apparent length measured by f. But, according to our criterion, this interpretation is inadmissible because the theory gives a strictly physical characterization of f, i.e. one that contains no psychological and sociological concepts. In sum, the referents of SR are b, f, and w, all three physical objects. Hence realism is upheld in relation to special relativity.

Example 2 Look at Heisenberg's indeterminacy formula:

$$\Delta_{\psi(\sigma, \sigma')}p(\sigma) \cdot \Delta_{\psi(\sigma, \sigma)}p(\sigma) \geqslant h/4\pi \qquad \text{(QM)}$$

If a correct deduction from first principles is performed and analyzed, it is seen that $\psi(\sigma, \sigma')$ represents the state of a microsystem σ (e.g. an iron atom) in an environment σ' (e.g. a magnetic field) at a given instant of time. It is further realized that $\Delta_{\psi(\sigma, \sigma')}p(\sigma)$ represents the width of the momentum distribution of the microsystem σ when the system-environment complex is in the state represented by $\psi(\sigma, \sigma')$. Similarly $\Delta_{\psi(\sigma, \sigma')}q(\sigma)$ stands for the width of the position distribution of the microsystem. In short the object variables or referents of the formula QM are σ and σ'. The phenomenalist claims that σ' is the observer-*cum*-apparatus complex. However, (*a*) the formula holds even in the absence of an environment, i.e. when σ' happens to be the null individual; (*b*) the formula holds whether or not σ' includes a measurement device – for example it holds for the atoms in a star; and (*c*) the theory does not specify the properties of either observers or measurement devices: it is a strictly physical theory and a thoroughly general one, married to no particular laboratory, let alone to a particular mind. Conclusion: the phenomenalist or operationist interpretation of quantum mechanics is groundless: as with the case of relativity, that interpretation is a superstructure that the formulas of the theory fail to support.

In conclusion: there are two main ways of looking at a scientific theory. One is through a semitransparent mirror, like operationism, that exhibits the superposed images of the referent and of ourselves. The second way is by using the non-reflecting magnifying lenses of axiomatics and semantics applied to the theory itself regardless of its philosophical wrapping. The second method alone guarantees the objectivity characteristic of science. (For details see Bunge, 1973a and 1973b.)

5. Relevance

5.1. *Kinds of Relevance*

The term 'relevance' is very much in vogue these days but it is seldom analyzed. In particular it is not always realized that relevance is a relation and that its relata can be of various kinds, as shown in the following diagram.

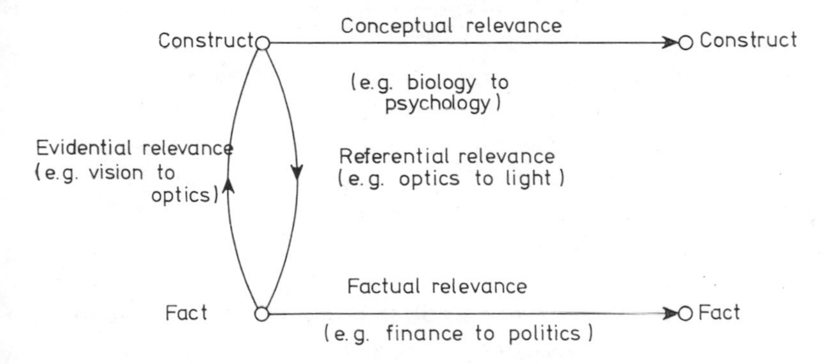

The relevance relations involved in the preceding diagram are these:

(i) construct-construct as regards form, i.e. *formal relevance*;
(ii) construct-construct as regards reference, or *semantical relevance*;
(iii) construct-fact, or *referential relevance*;
(iv) experiential fact-construct, or *methodological relevance*;
(v) construct-action, or *pragmatic relevance*;
(vi) fact-fact, or *factual relevance*.

The last two kinds of relevance may be characterized as follows. A fact may be said to be relevant to another fact if and only if the former makes some difference to the latter. And a construct may be regarded as being pragmatically relevant to an action iff the former is part of a view or theory that is instrumental in bringing forth or preventing the given action. We shall be concerned only with the first four types of relevance. They will be characterized in the following way.

DEFINITION 2.34 Let c and c' be two constructs. Then c is said to be *syntactically relevant* to c' iff there is a context in which c is logically related to c' in such a way that c determines c' at least in part.

Example 1 In a definition, the definiens is syntactically relevant to the definiendum. *Example 2* In an argument, the premises are syntactically relevant to the conclusions. *Example 3* In a function, the independent variables are syntactically relevant to the dependent variables. If the function has an inverse, the relevance is mutual.

DEFINITION 2.35 Let c and c' be constructs. Then c is said to be *semantically relevant* to c' iff (i) c is syntactically relevant to c' and (ii) c and c' share referents, i.e. $\mathcal{R}(c) \cap \mathcal{R}(c') \neq \emptyset$.

Example 1 The specific gravity function is relevant to the weight function. *Example 2* Let c concern a gene in an organism and c' a molar or phenotypical trait of the same organism. Then c will be semantically relevant to c' just in case genetics happens to contain a law according to which c determines c' at least in part. *Example 3* The biological variables are irrelevant to the psychological ones in the context of behaviorism.

DEFINITION 2.36 A construct c is said to be *referentially relevant* to a fact (thing, state, event, process) f iff f is in the reference class of c, i.e. if $f \in \mathcal{R}(c)$.

Example Pure mathematics is syntactially relevant to science, which is in turn referentially relevant to reality. *Example 2* According to the Copenhagen interpretation of the quantum theories, the latter are relevant to the human mind, while according to the realist interpretation they are not. *Example 3* The concept of thinking (or ideation) is referentially relevant to neural activity.

We may take a further step and introduce an absolute and a comparative concept of degree of referential relevance:

DEFINITION 2.37 The degree of *referential relevance* of a theory T to a domain F of facts is equal to $\mathscr{R}(T) \cap F$.

DEFINITION 2.38 Let T and T' be two theories that are referentially relevant to a domain F of facts. Then T is *referentially more relevant* to F than T' iff $(\mathscr{R}(T) \cap F) \supset (\mathscr{R}(T') \cap F)$.

Example Biology is referentially more relevant than ethics to the facts of animal aggression, altruism, and cooperation; and political science is more relevant to war than either biology or ethics.

These concepts should not be mixed up with the methodological concept of evidential relevance, that can be elucidated by

DEFINITION 2.39 An empirical fact e is *evidentially relevant* to a construct c iff there is another construct c' such that (i) c' is syntactically relevant to c and (ii) c' is referentially relevant to e.

Example 1 The conceptual basis of the common lie detector is the hypothesis that increase in hand sweating is an index of lying. *Example 2* Dreams are (so far) irrelevant to the study of personality because there is no scientific theory in which dream contents and personality traits are related. *Example 3* Prior to the theory of evolution the differences among species were not normally regarded as pointing to (or against) the hypothesis of evolution.

In Sec. 4.2 we elucidated the concept of semantic commensurability. We can now supplement this notion with the one of methodological commensurability, which has recently come to the fore with reference to scientific revolutions. We first stipulate

DEFINITION 2.40 Let T and T' be two factual theories. Then T and T' are said to be *methodologically commensurable* iff there are empirical facts that are evidentially relevant to both T and T'.

DEFINITION 2.41 Let T and T' be two methodologically commensurable theories. Then T is said to present a *larger methodological exposure* than T' iff the set of empirical facts evidentially relevant to T includes the one evidentially relevant to T'.

Example 1 Relativistic mechanics presents a greater methodological exposure than classical mechanics because it is inextricably related to electrodynamics. *Example 2* Molecular genetics presents a greater methodological target than Mendelian genetics because it is sensitive to an additional set of evidence, namely that of the molecular structure of genes. *Example 3* Any neuropsychological theory presents a greater methodological exposure (hence more risks) than the corresponding behaviorist theory (if any) because it can claim both behavioral and neurophysiological evidence pro and con.

Now, testing a theory presupposes knowing what the theory is about – not the other way around. In other words, contrary to the verification tenet of meaning, we hold that the semantics of a theory must precede its methodology (recall Sec. 2.6). More explicitly, we lay down the methodological

THESIS 2.4 If two scientific theories are methodologically commensurable then they are semantically (referentially) commensurable as well.

The converse is false. Hence referential and methodological commensurability are inequivalent. Thus two rival theories about ghosts are singly untestable, hence methodologically commensurable in the trivial sense that no fact of experience is relevant to either. The above principle contradicts the claim – originally put forth by Kuhn and Feyerabend – that every new revolutionary theory is methodologically incommensurable with its predecessor: that is is accepted for reasons other than having been shown to be truer. On the other hand our principle fits in with the actual practices of (*a*) comparing only semantically commensurable theories, (*b*) looking for evidence relevant to the reference class shared by the rival theories, and (*c*) making rational (though perhaps wrong) choices among rival theories after having weighed their respective empirical (predictive) performances. This does not entail that factworthiness is all-important: but it is necessary.

In examining the performance and the value of alternative factual theories it is usually taken for granted that they are semantically and methodologically commensurable. That is, one uses tacitly

DEFINITION 2.40 Two factual theories are *commensurable* iff they are referentially as well as evidentially commensurable.

A final convention: Two theories may be regarded as being *semantically rival*, or *competing*, iff, being commensurable, they have different senses, i.e. if they do not 'say' the same. (Examples: rival theories of learning, or of social mobility.) If two theories had the same sense then they could differ only in the organization of the material: in this case they would be just different formulations or presentations of one and the same body of knowledge. But the concept of sense lies far ahead.

Before applying the preceding ideas on relevance we note that they are relevant to the intuitive notion of a "category mistake" (Ryle, 1949). In the light of the preceding considerations it is clear that there are no absolute category mistakes: that whether or not a given association of predicates is mistaken depends on the theory that is adopted. It is equally clear that any such mistakes are epistemic not linguistic mishaps: they do not consist in violations of linguistic rules but in departures from accepted bodies of substantive knowledge. Thus while in the context of behaviorist psychology (adopted by Ryle) the coupling of mentalistic concepts to either behavioral or neurological ones would constitute an unforgivable category *faux pas*, in either a more backward or more advanced psychological context such predicate combinations might be correct. The science of the day, not either grammar or literary criticism, is competent to judge whether any given predicate association is right or wrong. In short, category mistakes are scientific mistakes.

5.2. *The Paradox of Confirmation as a Fallacy of Relevance*

Consider the empirical generalization ⌜All ravens are black⌝, or $\ulcorner(x)\,(Rx \Rightarrow Bx)\urcorner$ for short. This statement amounts to $\ulcorner(x)\,(\neg \mathrm{Rx} \vee \mathrm{Bx})\urcorner$ as well as to $\ulcorner(x)\,(\neg Bx \Rightarrow \neg Rx)\urcorner$. At first blush, finding anything that is not a raven, such as a book, or anything black, such as a tunnel, or even that is not black, such as a carrot, would seem to count as positive evidence for the generalization. But then almost anything would so count no matter where we look. Hence experience will have little if any value: confirmation will be so cheap as to be worthless. This is one of the paradoxes of confirmation (Hempel, 1945).

This paradox, which occurs neither in real science nor in our semantics, illustrates the fallacy of irrelevance. Indeed the statement that all ravens are black happen to concern birds not anything else: this is how ornithologists understand it. Therefore to put it to the test the ornithologist will scan only the *relevant evidence*, i.e. data about birds, without bothering about nonbirds, whether black or colored. This procedure is consecrated by our semantics, in particular by Definition 2.9 (iii) and Corollary 2.4b. Indeed, the reference class of the universal conditional ⌜$(x)\,(Rx \Rightarrow Bx)$⌝ equals the reference class of the molecular predicate $R \Rightarrow B$. And this predicate is defined on the intersection of the domains of its components R and B, as stipulated in Ch. 1, Sec. 2.3. Now,

$$\text{Dom}\ R = \text{Birds}, \quad \text{and} \quad \text{Dom}\ B = \text{Things}$$

hence

$$\mathscr{R}_s\,(\text{All ravens are black}) = \text{Birds} \cap \text{Things} = \text{Birds}$$
$$\mathscr{R}_s\,(\text{Anything is either not a raven or black}) = \mathscr{R}_p\,(\neg R \vee B)$$
$$= \mathscr{R}_p\,(R \vee B) = \text{Birds}$$
$$\mathscr{R}_s\,(\text{All nonblacks are nonravens}) = \mathscr{R}_p\,(\neg B \Rightarrow \neg R)$$
$$= \mathscr{R}_p\,(B \Rightarrow R) = \text{Birds}.$$

Therefore only an examination of birds will be relevant to the given generalization and thus be in a position to confirm or infirm it. (The discovery of albino ravens has refuted the generalization anyway.) *Moral 1*: If semantics is to be relevant to actual science it must be consonant with the methodology of science. *Moral 2*: It is not helpful to jump from the syntax (or logic) of science to its pragmatics (or methodology) without stopping over its semantics.

Our analysis of the empirical generalization about ravens carries over to statements in theoretical science. A typical example of a simple conditional occurring in the latter is ⌜All metals are conductors⌝, or ⌜$(x)\,(Mx \Rightarrow Cx)$⌝ for short. This is a law proper for, in addition to being well corroborated, it belongs to a definite scientific theory, namely solid state theory. (For the underlying criterion of lawfulness cf. Bunge, 1967a, Vol. I, Ch. 6, Sec. 6.6.) This theory accounts for conductivity (both thermal and electrical) in terms of metal structure: it explains why metals should be conductors. In the terminology of the preceding section: M is semantically relevant to C because there is a definite context, namely solid state theory, where (*a*) M and C have common referents and (*b*) M determines C. On the other hand in ⌜All ravens are black⌝ the predicate "is a raven" is so far semantically irrelevant to "is black": at the time of writing there seems to be no confirmed scientific theory explaining the blackness of ravens in terms of any of their basic properties. In other words, in the context of present day ornithology "raven" and "black" are syntactically irrelevant to each other, and *a fortiori* semantically irrelevant as well. This situation, atypical of science, is typical of ordinary knowledge: it suggests Hume's and James' ontology of universal ontic irrelevance. Such a metaphysics can hardly avoid fallacies of irrelevance. Not so the ontology of science, which is a web of laws. But this is a different matter. In any case our semantics may be of help when the metaphysician goes astray.

6. Conclusion

The notion of reference, though central to the semantics of factual science, is of little interest to logicians and mathematicians, whence it has been neglected by semanticists. The only serious attempt to analyze reference in exact terms seems to have been Goodman's. Unfortunately the definition he offered was exceedingly complex and at the same time it did not require predicates to be analyzed as functions from individuals to statements. Consider: "S is *absolutely about* k if and only if some statement T follows from S differentially with respect to k". In turn "a statement T follows from S differentially with respect to k if T contains an expression designating k and follows logically from S, while no generalization of T with respect to any part of that expression also follows logically from S" (Goodman, 1961, p. 7). The notion of aboutness is thus made to depend upon the concepts of consequence and denotation, but the latter is left unanalyzed. If only for this reason the elucidans turns out to be more obscure than the elucidatum (Patton, 1965). No wonder Goodman's criterion has not been applied to solve any of the genuine problems of referential ambiguity listed in Sec. 1.

Our approach to the problem of reference has been direct: we have been concerned with solving the problems mentioned in Sec. 1 and with proposing a solution that could be checked. Our proposed solution has consisted, in a nutshell, in bringing to light the domain of the predicate of interest. Since the responsibility for the characterization of a factual predicate rests ultimately with a scientific theory, it is to the latter that we should turn for an identification of the entities it purports to be about. Mark: to the theories themselves rather than to any extrasystematic ("philosophical") comments about them – the more so since such comments are often philosophically biased. If somebody claims that a theory T is about entities of kind K, let him formulate T in such a way that K occurs in the domain of at least one predicate in T: otherwise the claim will be groundless. But even showing that a theory refers to entities of some kind does not prove that it describes or represents them. For example, a theory of the effects of food deprivation in children will refer to food without describing or representing it. But the concept of representation deserves a separate chapter: the next.

CHAPTER 3

REPRESENTATION

According to a realist semantics, scientific theories represent their referents. They constitute *conceptual representations* of actual or hypothetical bits of reality – or rather of some of their traits. The same holds for some of the predicates and some of the formulas of a scientific theory. Just for some: not all of them represent. Thus, while a distance function may represent the spacing of things, not every one of the infinitely many functions of a given distance function will represent something. Likewise not every line in a deduction within a scientific theory represents some aspect of a thing: some statements are purely mathematical. What makes some constructs representational and others not? This is, in a nutshell, the problem of the present chapter.

1. Conceptual representation

There is an analogy between scientific theories and pictures. In both cases the represented object may be given (rather than invented) and the representation may be more or less accurate or truthful. This analogy is the basis for the expressions 'Scientific theories depict (or portray) their referents' and 'Science mirrors (or reflects) reality'. However, these are just metaphors, hence they may not go to the heart of the matter. Indeed, the manner of representing differs in either case. First, while a pictorial representation is itself a physical object, a conceptual representation is a thing of reason. Second, while in the former case only appearances can be pictured and all else can at best be hinted at, scientific representations do not pause at the skin of reality: they aim at representing the real, which is mostly hidden to the senses and alien to ordinary experience. Third and consequently, scientific representations are symbolic (yet not metaphoric) rather than pictorial even though they may contain a few pictorial ingredients. Fourth, while a picture is bound to be interpreted (and often also felt) in as many ways as there are beholders in different moods, a scientific representation is supposed to be objective. Fifth, while the aim of the arts is to arouse or to lull, to entertain or to edify – in any event to play up on our emotions – the aim of a scientific representation is to describe its referents in a truthful way. Whatever aesthetic emotions a scientific representation may arouse derive from realizing its logical and methodological virtues or shortcomings.

In view of these important differences between pictorial representations and conceptual representations, it seems advisable to speak of *conceptual reconstructions* rather than *images* of reality. Reconstructions are constructions, artifacts resulting from hard and often ingenious work, not just impressions and images – which can be had for free. Flies have an image of reality and so have scientists, but the latter possess, in addition, what they create, namely the conceptual representations of objects that cannot be grasped with the senses. These representations are only partial and at best approximately true but they can be checked and either improved or replaced by truer ones. What is more, unlike images they are built with mathematical bricks: one more reason for not calling them 'pictures'. More on the issues of critical realism in Ch. 10, Sec. 3.3.

Not every construct represents. To begin with logical concepts are nonrepresenting even when they are referential. Thus on our theory of reference (Ch. 2, Sec. 3.2) disjunction concerns propositions. But it represents nothing. And analytic statements may refer to anything but they describe or represent nothing except logical objects: logic does not mirror reality but rather the structure of human knowledge (cf. Bunge, 1974a). Likewise with other formal objects: they too are nonrepresentational. For example the number 8 fails to represent anything. True, 8 happens to be the number of genuine planets in our solar system and thus looks as if it did represent a property of a concrete thing. Not so: 8 is not the property of any one aggregate although it may be construed as the class of all octuplets. This is why it may pop up in any context, whether formal or factual. In other words, although the number of planets equals 8,8 is not identical with the number of planets. (That is, the relation of equality involved in ⌜The number of planets equals 8⌝ is not symmetrical, hence it cannot be the one of identity. Consequently instead of writing 'Card (set of planets) = 8' we should adopt the Algol convention and write 'Card (set of planets) := 8'. Likewise to indicate that we assign x the value 8, or that we set x equal (not identical) to 8, but not *vice versa*, we should write '$x := 8$' instead of '$x = 8$'.) In any event formal constructs need not, and often cannot, represent anything – except perhaps further formal constructs, as when a point in a manifold is represented by an n-tuple of numbers, or when a function is represented by a series.

A neutral construct, such as a set, may represent some concrete thing or aggregate of things. Thus a portion of a continuum may be made to represent a body, and a graph may be assumed to represent an institution. But such constructs stand on their own feet and are portable from one field of inquiry to another. Hence it is not by analyzing them that we may discover what if anything they represent. This can be found out only by examining the role such constructs play in scientific theories. Such roles are sometimes assigned explicitly, namely by semantic assumptions, as will be seen in Sec. 4. In any case reference and representation are independent since nonreferential constructs, such as sets, may be made to represent while referential constructs, such as tautologies, may not represent anything. In other words it is not the case that whatever represents refers and conversely. What is true is that scientific theories both refer and represent.

Moreover not every factual statement, i.e. one with factual referents, represents facts. Consider the irreducibly negative propositions. We shall call a statement *irreducibly negative* if it cannot be transformed into an equivalent positive statement except by the trick of introducing negative or disjunctive predicates – none of which can represent properties of real things. Thus ⌜Snow is not blue⌝ is irreducibly negative although it can be converted into ⌜Snow is nonblue⌝, which is grammatically affirmative but semantically negative, since "nonblue" represents no color. Likewise ⌜There are no ghosts⌝ can be transformed into ⌜Everything is nonghostly⌝, but since "nonghostly" corresponds to no property that can be pinned down as being possessed by a thing, the statement is irreducibly negative. Our two examples illustrate a class of statements with factual reference and moreover true, but that fail to represent any facts. In general: If a proposition p represents a fact f, then $\neg p$ is just the denial of p, not a representative of non-f. (See Kraft, 1970.) In other words there are no negative facts: negation is a purely conceptual operation without an ontic correlate. We must resist the attempt to save the view that language is a picture of reality by introducing the fiction of the negative fact (Russell, 1918). The number of facts cannot be duplicated by just issuing a semantic decree. In conclusion: irreducibly negative statements do not represent anything even when true. Hence not everything that refers represents.

On the other hand every positive statement does constitute a partial representation of its referents. In particular a positive factual statement represents a fact or rather some facet of it. Thus ⌜*b* grows faster than *c*⌝ refers to *b* and *c* and represents (truly or falsely) the fact that *b* grows faster than *c*. The negate of the same proposition has the same reference class, namely $\{b, c\}$, but does not represent the "negative" fact that *b* fails to grow faster than *c*: it is just the denial of the former statement.

The distinction between reference and representation is no idle philosophical technicality: it is relevant to our understanding of topical scientific controversies. For example, biologists are still arguing about the authentic referents of the synthetic (neodarwinian) theory of evolution. So far they have produced no conclusive argument for either of the theses in dispute: that the theory is concerned with individual organisms, or with populations, or with species. However, a semantic glance at the typical formulas of the mathematical theory of evolution shows the following. First, the theory *refers* to populations or aggregates of coexisting and interacting members of a given kind (species). Hence it uses all three concepts: those of individual, species, and population. Second the theory *represents* both individual and collective traits – among other things the occasional changes in kind (speciation and extinction) occurring in a population. The failure to realize that there is no incompatibility among the three concepts, because they perform different and complementary roles, may be blamed not only on the obsolete dichotomy Platonism-nominalism, but also on the backwardness of semantics, which has never helped science find its way.

The difference between reference and representation becomes particularly clear in advanced theories such as those in physics. For example, here a probability function will *refer* to some system or some state(s) of it, while the values of that function may be taken to *represent* certain dispositions of the system – much as the mass function M refers to bodies while a particular value $M(c)$ of M will represent the mass of the body c. In quantum mechanics, every dynamical property of a system, such as its linear momentum, is represented by some operator in a Hilbert space. That is, the operator represents a property of its referent. In statistical mechanics the partition function of a multicomponent system refers to the latter but fails to represent any single property of it. It accomplishes much more: it generates the conceptual representatives of all the thermodynamic properties of the physical system. And in electromagnetic theory the value $E(\phi, x, t)$ of the vector valued function E at the field ϕ at the place x and the instant t, represents the strength of the electric component of the referent ϕ at x and t. There are infinitely many constructs in the theory, such as the powers and derivatives of E, all of them with the same referent but which represent no trait of it.

To sum up. While in factual contexts the reference relation pairs a construct off to a thing as a whole, or to a collection of things, the representation relation matches a construct with some aspect or property of the thing or collection of things. Just as we read '$\mathscr{R}cf$' as 'c refers to f', so we may abbreviate '*c represents f*' to '$c \hat{=} f$'. If c happens to be a quantitative construct we may also read '$c \hat{=} f$' as 'c *represents* the strength of f', where f is a property not a whole thing or a whole fact. For example, in mathematical neurobiology the element a_{mn} of a certain matrix is assumed to represent the strength of the action (excitatory or inhibitory) of neuron m on neuron n. Table 3.1 exhibits some typical examples of conceptual representation that will guide our subsequent investigation.

2. THE REPRESENTATION RELATION

2.1. *A Characterization*

The relation $\hat{=}$ of conceptual representation pairs some constructs off to some objects, whether conceptual or factual. We shall restrict our study to the case when the represented object is factual, i.e. when it is a thing or an aggregate of things, a property of either or a change in one or more properties of a system – i.e. an event. The main types of representative and their respective representees are shown in Table 3.2.

A few comments are in order. Firstly, we have included no individual constants in our list. And this because they can denote but not represent. Thus the man Socrates is denoted by his name but is represented only by certain propositions about him as well as by some definite descriptions of him. And an unspecified individual thing (an arbitrary element of a set of things) is likewise denoted by an individual variable but not represented by it. Only a collection of statements can represent an unspecified individual thing. Individual things, whether specific or unspec-

TABLE 3.1

Constructs: representing and nonrepresenting

	Construct	Represents	Refers to
Nonreferential concepts	3	————	————
	Open set of R^n	Open set of an n-manifold	————
	Region of a 3-manifold	Body or force field	————
Referential concepts	$Q(b)$	Charge of b	Body b
	$P(s)$	Probability of s	System in state s
	$6n$ dimensional manifold	Dynamical states of a system with n components	System with n components
	Partition function	————	Multicomponent system
Statements	The lake froze.	A freezing of the lake	The lake
	The lake felt cold.	Joint property of lake and subject	Lake and subject
	There are no green people.	————	People
Theories	Maxwell's field theory	Structure and spreading of electromagnetic fields	Electromagnetic fields
	Theory of evolution	Emergence, evolution and extinction of populations belonging to different species	Populations or organisms
	Theory of social mobility	Changes in occupation, social class, income bracket, or dwelling place	Groups of people

TABLE 3.2

What represents what

Representing construct	Represented object
Set of statements (e.g. theory)	System (individual or aggregate)
Predicate or structure (e.g. a set together with a relation on it)	Property of a system, relation or connection among systems
Set of singular or existential statements	Fact involving one or more systems (state, circumstance, or event)
Set of universal statements (e.g. law statements)	Pattern of a system's composition, structure, or change

ified, are best represented by logically organized sets of statements, i.e. by theories.

Secondly and as a consequence: sets devoid of structure do not represent either. Thus we may stipulate that 'E' denotes the population of African elephants, but E is incapable of representing anything. Only a structured set, such as a set together with certain functions, or a family of subsets of a given set, can represent something. And structured sets are structures, such as a semigroup or a topological space.

Thirdly, we have included existential statements among the representing constructs. It might be objected that a proposition such as ⌜There are flies⌝ hardly represents a fact. Yet it does represent a condition of the world, albeit in an imprecise fashion. ⌜There are flies in this room⌝ is a more precise statement – quite apart from its truth value. A whole sequence of such statements, each more precise than its predecessor, can be imagined. They will have different contents but every one of them, even the first statement of unqualified existence, may be regarded as representing. In sum an existential statement, if referring to some things, does constitute a partial albeit hazy representation of some facts involving those things.

Fourthly, note that we distinguish between a law statement and that which such a proposition is supposed to represent – an objective pattern. We distinguish, that is, between a law and a law statement. This distinction, emphasized by the physicist-philosopher Ampère and denied or overlooked by a number of philosophers (notably Hume, Kant, Peirce, Boutroux and their respective followers), is necessary to account for the fact that each objective pattern can be represented in a number of different ways, as will be seen in Sec. 2.2. It is also needed to explain both the process of theory construction and the history of science – which history may be regarded as a sequence of attempts to build improved conceptual representations of objective patterns or laws.

We summarize the preceding remarks in the following rather informal

DEFINITION 3.1 The relation $\hat{=}$ of *factual representation* is a relation from constructs to facts (i.e. such that $\mathscr{E}(\hat{=})\subset C\times F$) subject to the following conditions:

(i) the properties of real things (including their interactions with other things) are represented by predicates (in particular functions);

(ii) real things are represented by sets structured by relations, functions, or operations;

(iii) facts (e.g. events) are represented by sets of singular or existential statements;

(iv) the stable (recurrent and invariant) patterns of the constitution and behavior of real things are represented by sets of universal statements.

Aside from this, as here construed $\hat{=}$ has no simple properties. In particular it is not symmetric: the represented objects do not represent back their representatives. Consequently $\hat{=}$ is not reflexive: constructs do not represent themselves. And the question whether $\hat{=}$ is transitive is pointless, since facts do not represent anything.

When the representing construct is a theory (hypothetico-deductive system) we try and be more precise, speaking of a representation function rather than just relation. In fact we propose

DEFINITION 3.2 Let T be a theory about entities of kind K and call $S(x)$ the collection of possible states (i.e. the state space) of thing $x \in K$. Further, call $S = \bigcup_{x \in K} S(x)$ the union of the state spaces of all the members of K (so that an arbitrary state of an arbitrary element of K is in S). Then T is said to be a *representation* of K's iff there exists a function $\hat{=} : S \rightarrow T$ assigning to every state $s \in S$ a statement $t \in T$. In this case '$t = \hat{=}(s)$' is read 't represents s'.

Now, a representation may be poor, fair or good, depending upon how complete and true it is. The ideal is characterized by

DEFINITION 3.3 Let $\hat{=} : S \rightarrow T$ be a representation of things of some kind. Then $\hat{=}$ is said to be *accurate* iff (i) it is bijective and (ii) $\hat{=}(s)$ is true for every $s \in S$.

As a matter of fact there are no accurate theoretical representations of real systems: every theory misses some possible states, or it includes some impossible states, or the formulas of it are only approximately true. Nevertheless one often reads that probability does constitute an isomorphic and true representation of chance events. The argument usually runs like this. Consider the possible outcomes of a random experiment such as the simultaneous flipping of two coins. Form the set E of all possible outcomes, such as getting at least one head, or not getting any head in the trial. It is easily seen that E has a Boolean structure. Now form the set T of statements describing such possible events. This set, too, will be a Boolean algebra. Hence there is a bijection $\hat{=}$ that maps E on to T in such a way that, if e_1 and e_2 are in E, then $\hat{=}(e_1)=t_1 \in T$, $\hat{=}(e_2)=t_2 \in T$, $\hat{=}(e_1 \cap e_2)=\hat{=}(e_1) \,\&\, \hat{=}(e_2)$, $\hat{=}(e_1 \cup e_2)=\hat{=}(e_1) \vee \hat{=}(e_2)$, and finally $\hat{=}(\bar{e})=\neg\hat{=}(e)$. The weakness of this reasoning is, of course, that E is not a set of actual facts but of possible facts: actual events are "positive" and "definite" (simple or composite but never alternative). Hence there is no bijection mapping actual chance events on to a theory. What is true is, that probability theory is part of the formal background of any stochastic theory furnishing a (never fully accurate) representation of some factual domain.

We postpone a fuller discussion of $\hat{=}$ to Ch. 6, Sec. 3. Here we note the relations between $\hat{=}$ and other semantic relations, as summarized in Figure 3.1.

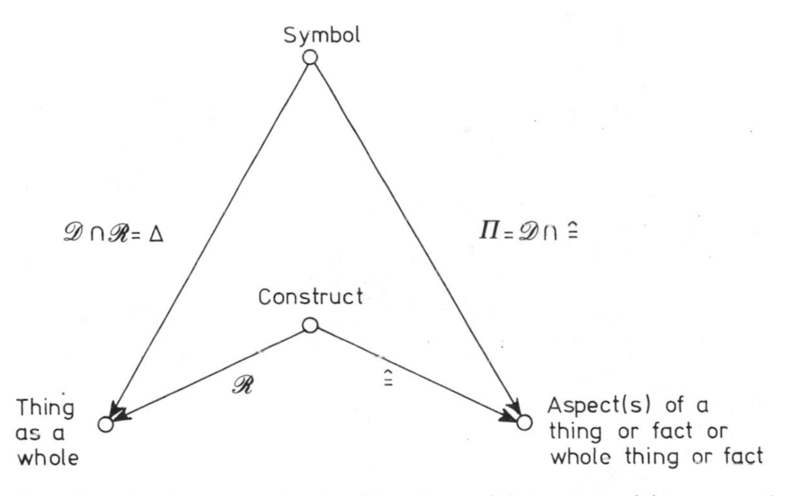

Fig. 3.1. Relations between designation (D), reference ($\mathscr{R}$), denotation (Δ), representation ($\hat{=}$), and proxying (Π).

The new relation occurring in the diagram, i.e. the intersection Π of designation $\mathscr{D}$ and representation $\hat{=}$, can be interpreted as the relation of *proxying* or deputizing. Thus we may say that, in a certain context, the symbol 'V' designates the (or rather a) velocity concept, which in turn

TABLE 3.3

Theory of the vibrating string: its main constructs and what they represent

Construct	Represents	Status
x	Projection of arbitrary point of string on Ox axis	Basic concept
l	Length of projection of string on Ox axis	Basic concept
t	Arbitrary instant of time	Basic concept
$u(x, t)$	Vertical displacement of arbitrary point of string at time t	Basic concept
T	String tension	Basic concept
ρ	String mass density	Basic concept
$(T/\rho)^{1/2}$	Velocity of travelling waves	Defined concept
$\frac{1}{2}\rho(\partial u/\partial t)^2\,\mathrm{d}x$	Kinetic energy of element $\mathrm{d}x$ of string	Defined concept
$\frac{1}{2}T(\partial u/\partial x)^2\,\mathrm{d}x$	Potential energy of element $\mathrm{d}x$ of string	Defined concept
$E =_{df} \frac{1}{2}\int_0^l \mathrm{d}x\,[\rho(\partial u/\partial t)^2 - T(\partial u/\partial x)^2]$	Total energy of string	Defined concept
$\mathscr{L} =_{df} \frac{1}{2}[\rho(\partial u/\partial t)^2 - T(\partial u/\partial x)^2]$	Excess of kinetic over potential energy densities (lagrangian density)	Defined concept
$\int_{t_1}^{t_2} \mathrm{d}t \int_0^l \mathrm{d}x\, \mathscr{L} =$ maximum or minimum	Pattern of all vibrations of string as a whole over an arbitrary time interval $[t_1, t_2]$ (global law)	Basic law statement
$T(\partial^2 u/\partial x^2) - \rho(\partial^2 u/\partial t^2) = 0$	Pattern of all vibrations of an arbitrary point of string at an arbitrary instant (local law)	Derived law statement
$u_1 = \phi_1(x - Vt)$	Wave travelling along Ox axis from left to right with speed V	Derived law statement schema (arbitrary ϕ_1)
$u_2 = \phi_2(x + Vt)$	Wave travelling along Ox axis from right to left with speed V	Derived law statement schema (arbitrary ϕ_2)
$E = \mathrm{const}$	Total energy of string remains constant in course of time	Derived law statement
$u(x, 0) = f(x)$ (Initial conditions with f and g given)	Initial shape of string	Subsidiary hypothesis or datum
$(\partial u/\partial t)(x, t) = g(x)$ (Initial conditions with f and g given)	Initial velocity of arbitrary point of string	Subsidiary hypothesis or datum
$u(0, t) = u(l, t) = 0$ (boundary conditions)	Ends of string fixed at all times	Subsidiary hypothesis or datum

represents speed. But we may also say that (in the same context) 'V' stands for, or proxies, speed.

We close this section with Table 3.3, that exhibits the many items of one of the simplest and most characteristic of specific scientific theories. It shows clearly that the representation assumptions are an integral part of the theory: without them the latter reduces to a mathematical formalism.

2.2. *The Multiplicity of Representations*

Representations are not unique: one and the same factual item may be represented in alternative ways. Some such alternatives are equivalent, others not. For example, a region of physical space S may be represented as a subset of a certain manifold M, which may in turn be mapped onto some subset of the collection R of triples of real numbers. Hence physical space can be represented either as a portion of M or as a portion of R^3: see Figure 3.2. These two representations are inequivalent. But in turn

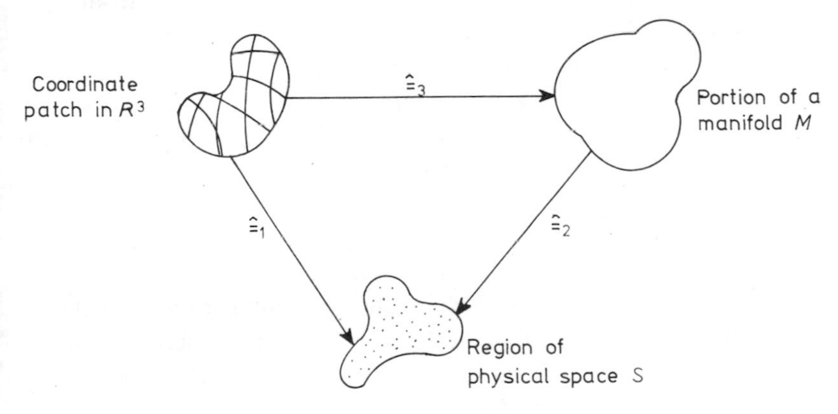

Fig. 3.2. The coordinate patch constitutes a construct representing another construct, namely a region of a manifold. And each of them constitutes a *sui generis* representation of a region of physical space.

every region of the number space may be mapped onto some other region of the same space by means of a coordinate transformation. Since there are infinitely many possible coordinate transformations, there are infinitely many possible representations of a given region of the manifold M, hence just as many numerical representations of the original region of physical space. The latter representations, i.e. the various coordinate patches mirroring one and the same region of M (or of S), are mutually equivalent. Consequently the choice among them is a matter of convenience not of truth. (Hence if meaning were dependent on truth, as conventional semantics has it, coordinate transformations should be pronounced meaningless.)

To compare alternative representations we need, at the very least, definite criteria for deciding whether any two given representing constructs represent the same factual item in different though equivalent ways. Physics abounds in such criteria, which are of a considerable epistemological and methodological interest. *Example 1* In classical mechanics any two solutions of the equations of motion, referring to the same system but to different reference frames, are equivalent representations of the same state of motion of the system provided they can be converted into one another by a Galilei transformation. *Example 2* In the relativistic theory of gravitation it is postulated that two solutions of the field equations referring to the same field and which can be transformed into one another by a continuous coordinate transformation represent the same state of the field. *Example 3* In the theory of the spinning electron every component of the electron spin is represented by an operator that is in turn representable by alternative matrices. Two such matrices represent the same spin component provided there exists a unitary transformation that carries one into the other.

Every one of the preceding criteria of equivalent representation has a definite place in some theory or other: there seems to be no theory free criterion. In any case the definitions we shall presently propose are theory bound. The first of them will concern alternative constructs in a theory, the second will handle the translation code relating mutually equivalent representations, and the third will concern equivalent theories.

The first of our forthcoming definitions hinges on the concept of basic law statement. This is a metascientific rather than a semantic concept, but we owe no apology for such an intrusion: it is unavoidable if our semantic theory is to be relevant to science. In any case the concept in question can be elucidated in the philosophy of science (Bunge, 1967a, Ch. 6). A hypothesis is called a *law statement* if and only if (i) it is universal in some respect (rather than restricted to a finite number of cases), (ii) it is systematic, i.e. a member of some hypothetico-deductive system, and (iii) it has been corroborated in some domain by scientific methods. And a proposition of this kind is called a *basic* (or *fundamental*) *law statement* of a theory T if and only if it derives from no other statement in T. We are now ready to state

DEFINITION 3.4 Let c and c' be two representing constructs belonging to a factual theory T. Then c and c' are *equivalent representations* of the same factual item [state, event or process] if and only if they can be freely substituted (i.e. substituted *salva significatione* and *salva veritate*) for one another in all the basic law statements of T, i.e. if the latter are invariant under the exchange of c for c'.

Example 1 The choice of different coordinate systems ensues in different representations of physical quantities. These representations are equivalent if they satisfy the same laws of motion and field laws. *Example 2* Let P and Q be two quantum mechanical operators representing dynamical variables. They will represent the same property of a physical system provided there exists a similarity transformation S between them, i.e. iff $Q = SPS^{-1}$. Proof: Similarity transformations leave operator equations invariant. For example, $P^2 + P + I = 0$, where I stands for the identity operator, goes over into $SPS^{-1}SPS^{-1} + SPS^{-1} + SIS^{-1} = 0$. Calling $SPS^{-1} = Q$, we retrieve the original statement in a different notation, i.e., $Q^2 + Q + I = 0$. *Example 3* Let the hamiltonian of a two component system be $H = H_1 + H_2 + H_{12}$. Perform a canonical (unitary) transformation of the Q's and P's that "eliminates" the unperturbed energies, i.e. that makes H collapse into H_{12}. The new variables constitute the so called 'interaction representation' of the system.

Definition 4 bears on a basic aspect of concept formation in factual science, namely the representation of properties of concrete systems. Chief among such representatives are physical magnitudes such as force, stress, concentration, and temperature. Every magnitude or quantity is representable by at least one function whose values depend not only upon the physical system itself but perhaps also on the conventional units agreed upon. In other words, magnitudes and their corresponding units (a whole class of them per magnitude) must be introduced at one stroke: the set of possible units for any given magnitude should occur in the very domain of "definition" of the function. For example, in elementary electrostatics the electric force F between any two point charges is a certain function $F: B \times B \times U_F \to R^3$, where B is the set of bodies, U_F the set of force units, and R^3 the set of ordered triples of real numbers. There are infinitely many units in U_F. Every choice among them will ensue in one force value. Similarly for every other magnitude provided it is endowed with a dimension.

A way of solving this problem in a general manner is by adopting, for every scalar quantity and every component of a vector or a tensor valued quantity, the following assumption (Bunge 1971a):

AXIOM Let $A, B, \ldots, N$ be n kinds of physical system endowed with a (mutual or joint) property P. And let R designate the real number system and $\mathscr{P}(R)$ the power set of R, i.e. the family of all real number intervals. Then for every property P there exists a nonempty set U_M, called the set of *M-units*, and there is at least one function

$$M: A \times B \times \cdots \times N \times U_M \to V, \quad \text{with} \quad V \subseteq R \quad \text{or} \quad V \subseteq \mathscr{P}(R)$$

called a *magnitude*, such that M represents P.

At first sight this assumption is redundant since, for every domain D, there are infinitely many functions "defined" on D with values in R or in $\mathscr{P}(R)$. (There are R^D such functions in the first case and $(\mathscr{P}(R))^D$ in the second.) But our axiom is not concerned with such functions except in their representative capacity. It postulates that, given any property of a system, whether known or unknown, there is at least one function that will *represent* it, i.e. that will satisfy the law statements that characterize the system. This postulate is almost in the nature of a hope: there might well exist a property not so representable. But then we would not know of its existence, for our knowledge of things and their properties *consists in* our representations of them. Furthermore the postulate does not assert that the representation of every property is unique: realistically enough the existential quantifier in it is indefinite. Different theories are likely to represent differently one and the same property. It is not up to semantics to decide which is the best representation of a given property: this is a task for science. What philosophy can do is to render explicit and systematize the criteria at work in the choice among the possible representations of a given property. One such (methodological) criterion is this: Given a property of a complex system, the best representation of it will be the one occurring in the truest and most numerous law statements concerning the given system.

2.3. *Transformation Formulas and Equivalent Theories*

To go back to the equivalent representations of a given set of factual items: how are those representations related? The answer is supplied by

DEFINITION 3.5 A statement in a theory T is called a *transformation formula* of T iff the statement relates equivalent representations of the same factual items [in accordance with Definition 4].

Example 1 The Lorentz transformation formulas relate the spatiotemporal coordinates of one and the same physical system in relation to equivalent frames of reference: they are transformation formulas of special relativity physics. *Example 2* The canonical (or contact) transformations relate different representations of the generalized coordinates and momenta of a system. Since they leave the basic (canonical) equations invariant, they are among the transformation formulas of any canonical theory.

Some remarks are in order. First, transformation formulas are not law statements and they are not data either, even though they belong in every factual theory containing spatiotemporal concepts. They just relate different representations. This point, which is obvious in the light of the preceding discussion, is very often misunderstood. Thus the Lorentz (and also the Galilei) transformation formulas are often interpreted as representing the switching on of a uniform motion – which of course would correspond to an acceleration not to a uniform relative motion. And at one time the quantal theory of canonical (unitary) transformations was regarded as the very kernel of the physical theory, whereas in fact it has no factual content of its own, being just the collection of bridges among different but equivalent representations.

We are now in a position to elucidate the notion of equivalent theoretical representation, which came to prominence in Galilei's trial and has been at the center of the realism vs. instrumentalism dispute ever since. (In fact one of the thesis of Cardinal Belarmino was that Galilei erred in holding that the heliocentric "system of the world" was true to fact while the geocentric system was false: he should have said instead that the two systems were equivalent. The same view has been advocated in recent times by conventionalists like Poincaré and by positivists like Frank and Reichenbach. However, the very notion of theory equivalence was never too clear in these debates.) We propose the following elucidation:

DEFINITION 3.6 Let T and T' be two theories with the same factual referents. Call $\mathbb{P}$ and $\mathbb{P}'$ their respective predicate bases. Then T and T' are said to be *semantically equivalent* (or to constitute *equivalent representations* of their referents) if and only if there exists a set of transformation formulas for $\mathbb{P}$ and $\mathbb{P}'$ that effects the conversion of T into T' and *vice versa*.

Example 1 Lagrangian and hamiltonian dynamics are equivalent representations of systems in general even though their formalisms are different. Indeed, there is a bridge or transformation formula between the two theories, namely $H=p\dot{q}-L$, that leaves the content invariant. *Example 2* On the other hand the geocentric and the heliocentric "systems of the world" are not semantically equivalent if only because the former has no equations of motion (but only equations for the trajectories). Only the planet trajectories, when written in geocentric or in heliocentric coordinates, are equivalent representations. Since such trajectories is all one can observe, an empiricist must conclude to the overall equivalence of the two representations. However, they are really different in every other respect: "factual" and "empirical" are not identical concepts. For example the Copernicus-Kepler-Newton representation of the solar system refers not only to the bodies in the system but also to the gravitational field that keeps them together, which the Ptolemaic representation did not. (Cf. Bunge, 1961c.)

We close this section by laying down some tenets that are germane to the preceding considerations although they belong in the pragmatics of science rather than in its semantics. Whichever their proper location here they come.

*P*1 For any factual item (thing, thing property, event) it is possible to build at least one construct that represents it.

*P*2 Given any representing construct it is possible to form at least one other construct that is semantically equivalent to the former.

*P*3 Given any representing construct it is possible to build a semantically stronger construct.

3. Modeling

3.1. *From Schema to Theory*

A single predicate, such as "round" or "competitive", can represent a trait of a complex system, never the total system. An adequate representation of a whole system, even if either is comparatively simple, requires a cluster of concepts – better, a whole theory, i.e. a body of logically interconnected statements. However, for ordinary life purposes as well as for restricted scientific purposes a mere list of salient properties is often sufficient. For example, "Brunette, medium height, 35-25-35, pretty, witty" can pass for a representation of a girl – actually a whole class of girls. Beyond this our conceptual representations of things actual or hypothetical come in all degrees of complexity and generality. It is convenient to distinguish the following kinds of representation of a system, in order of increasing complexity and generality (cf. Bunge, 1973a and 1973b).

1 *Schema* or *model object* = List of outstanding properties of an object of a given species. *Example* A neutral pion is a particle with mass 135 MeV and half life 10^{-16} sec, that decays mostly into two gamma photons.

2 *Sketch* or *diagram* = Graph of the components of an object of a given species and their functions and relationships. *Example* Flow diagram of a factory.

3 *Theoretical model* or *specific theory* = Hypothetico-deductive system of statements representing some of the salient features of a thing of a given species. *Example* A stochastic learning model.

4 *Framework* or *generic theory* = Theory representing the features common to all things of a given genus. *Example* The theory of evolution.

To put it more briefly: a schema lists items; a sketch displays in outline the relationships among the items of a schema; a theoretical model spells out the sketch; and a generic theory is a theory free from specifics but convertible into a theoretical model (specific theory) upon being adjoined a schema or model object. All four constructs are supposed to represent some real thing but each of them is necessarily incomplete as well as, at best, fairly faithful (true). These two shortcomings of our conceptual representations of the world cannot be helped except little by little and in two ways. First, by multiplying the number of conceptual representations (e.g. theoretical models) of the same object, having each of them focus on a different aspect of it: that is, by varying the viewpoint. Second, by improving on each of these partial representations. Actually this is what happens: at any given moment we have a stack of snapshots of an object, and at successive moments we have different and thicker stacks. (Proviso: that research be continued.) But enough of metaphors.

The various kinds of conceptual representation can also be characterized as follows.

FRAMEWORK (GENERIC theory) T_G
Referents: objects g of genus G
Primitive base: $\mathscr{B}(T_G) = \langle G$, Representatives of generic basic properties of the g's$\rangle$
Axiom base: $\mathscr{A}(T_G) =$ Basic assumptions "defining" $\mathscr{B}(T_G)$

THEORETICAL MODEL (SPECIFIC THEORY) T_S
Referents: objects s in species S
Primitive base: $\mathscr{B}(T_S) = \langle S$, Representatives of basic properties of the S's$\rangle$
Axiom base: $\mathscr{A}(T_S) =$ Basic assumptions "defining" $\mathscr{B}(T_S)$

SKETCH (DIAGRAM) D_S
Referents: objects s in species S
Concepts $= \mathscr{B}(T_S)$
Hypotheses: The bare bones of $\mathscr{A}(T_S)$.

SCHEMA (MODEL OBJECT) M_S
$M_S = \mathscr{B}(T_S)$.

The relations and differences between the four kinds of conceptual representation are now clearer. They can be summarized as follows:

(i) Whereas a schema is just a bunch of concepts, a sketch is a structure such as an oriented graph. A sketch includes a schema.

(ii) There is no logical difference between a theoretical model and a generic theory: both are hypothetico-deductive systems. The difference resides in their respective reference classes, and is reflected in the greater specificity of the basic assumptions of a theoretical model. Briefly, whereas $\mathscr{R}(T_G) = G$, $\mathscr{R}(T_S) = S \subset G$.

(iii) $\mathscr{B}(T_S) = \mathscr{B}(T_G) \cup M_S$, i.e. $\mathscr{B}(T_G) = \bigcap_{S \subset G} \mathscr{B}(T_S)$.

(iv) $\mathscr{A}(T_S) = \mathscr{A}(T_G) \cup H_S$, where H_S is a set of assumptions representing the specific characteristics of the members of S as outlined by the corresponding sketch or diagram.

(v) Neither is a picture of its referent: conceptual representations are all symbolic. Even scientific diagrams are symbolic and can be replaced by sets of statements. No theory can possibly resemble its referents. Thus there is no analogy whatever between a field and the differential equations representing it. Even Bohr's pictorial representation of the atom symbolizes but a small part of Bohr's model: it leaves out the equations of motion, the quantization conditions, and the resulting jump equations.

Table 3.4 displays some examples of the four kinds of construct we have just characterized.

TABLE 3.4

Some examples of conceptual representation

Object	Schema	Sketch	Theoretical model	Generic theory
Coin flipping	Ideal coin = $\langle$Head, Tail$\rangle$	Random sequence of Heads and Tails	Theory of Bernoullian sequences	Probability theory
Hierarchically organized group	Partially ordered set	Dominance graph or dominance matrix	Theory of dominance	———
Prey-predator system with 2 components, e.g. foxes & rabbits	Instantaneous numbers of foxes and rabbits	Numbers of foxes and rabbits in successive generations	Theory of fox-rabbit economy	Volterra-Lotka theory of predator-prey system
Corpuscle attached by spring	Mass, position, velocity, elastic force	Hamiltonian of system	Theory of the harmonic oscillator	Particle mechanics
Deuteron	Proton and neutron	Potential well	Quantum mechanics of potential well	Quantum mechanics

3.2. *Problems of Modeling*

The conceptual representation of things raises a number of interesting problems, some semantical, some methodological, and some technical, i.e. tractable with the resources of a special science. We can only mention a few such problems.

1 Given two representations find out whether they are *equivalent* rep-

resentations of the same system. For example, find out whether these two network diagrams are equivalent. (They are: see Seshu and Reed, 1961, pp. 1–2.)

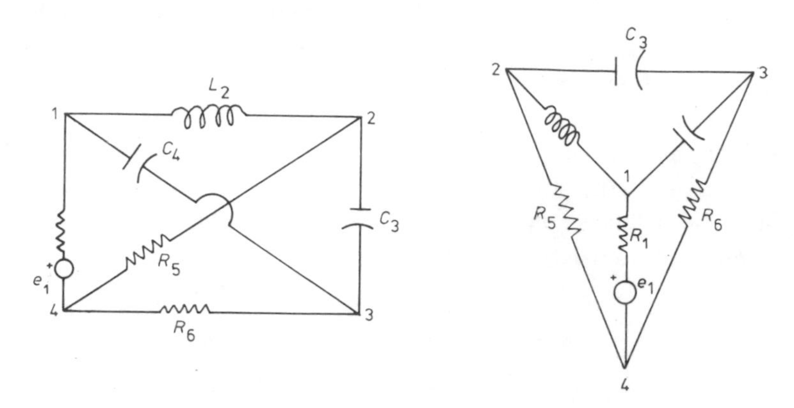

2 Given two inequivalent representations find out whether they concern *the same* system. For example, let Δ be a block diagram of an organism. This sketch is built with objects (e.g. sets) and mappings (e.g. functions) taken from a certain category A (e.g. the category of sets or the category of topological spaces). Now consider another category B and a functor F from category A to category B. Call $F(\Delta)$ the image of the original block diagram under the functor F. This image is an alternative block diagram of the same system just in case F satisfies rather restrictive conditions: if it is faithful, regular, and multiplicative (Rosen, 1958).

3 Given a real system decide which of its traits to chop off and which ones, going beyond the data at hand, to invent. The decision will depend on the goal no less than on the available information and on the conceptual tools accessible to the theoretician: there is no unique solution to this problem because there are no recipes for building theoretical models.

4 Given a real system decide which kind of representation to build: (*a*) a black box (exogenous variables only), (*b*) a grey box (both exogenous variables and internal states), or (*c*) a translucid box (both exogenous and endogenous variables, the latter representing features of the inner workings of the system). In this case an additional factor intervenes – namely the theoretician's philosophy. Thus while positivism favors the black

box, realism encourages translucid boxes – and spiritualism cares for neither.

Finally let us list a few topical problems in the current philosophy of science involving the concept of representation.

1 Is it necessary for a successful concrete representation of some object, i.e. for a simulate, to resemble the modeled object? Undoubtedly, for otherwise the simulate would not qualify as such: it could not possibly take the place of the original in any respect. However, resemblance is required only in the respects we want to imitate. For example, a blueprint of a house must respect the relative positions of the components of the house, and so must a ball and spoke model of a molecule. Other kinds of simulate of the same objects will seize on different features, such as behavior or gross output. For example a binary digital computer bears no detailed anatomical resemblance to a brain but its net behavior is in some sort of correspondence with the functioning of a brain engaged in algorithmic thinking.

2 Does every conceptual representation have to resemble its referents? No, it need not share any properties (substantial analogy) nor need it be a homomorphic image of the represented thing (formal analogy). For example, the theory of evolution does not resemble evolution.

3 Can a theoretical model represent every trait of each of its referents? Of course not. A basic procedure of scientific theorizing, in contrast to describing, is the discarding of details and idiosyncrasies, e.g. the treatment of equivalents as if they were identicals. In other words, it is a basic methodological principle of theoretical science that equivalents (or genidenticals) be represented as being identical.

4 Are theoretical entities real or fictitious? This is an ill conceived question. First, because the "entities" concerned are not entities at all but constructs occurring in scientific theories. Second, because they may not concern entities but properties of (hypothesized) entities, such as the energy of a collapsing star. The correct question is not whether "theoretical entities" are real but whether our theoretical concepts refer to real entities and, if so, which of them represent them correctly and, if so, to what extent. And these are hardly questions that can be answered with exclusively philosophical resources: they are matters for experimental science.

5 It has been claimed, without any supporting arguments, that "what

we have called the laws of nature are the laws of our methods of representing it. The laws themselves do not show anything about the world" (Watson, 1950, p. 52). True or false? Neither: just confused. The goal of natural science is to represent nature – an objective it achieves when finding its laws. The investigation of the patterns of representation, on the other hand, belongs to psychology, epistemology and methodology. An analysis of both laws statements and the concept of representation would have avoided the confusion.

To conclude: all science aims at producing conceptual representations of its referents. (See however MacKay (1969) for the claim that the making of representations is the concern of information theory.) In the process of building such representations scientists meet challenging methodological problems that cannot be appraised if hypothesizing, modeling and theorizing are conceived as data summarizing. Whereas some such problems are peculiar to the given field of research others are undisguised epistemological problems. Philosophy can help in understanding them or at least in realizing that they involve philosophical problems.

4. Semantic Components of a Scientific Theory

4.1. *Denotation Rules and Semantic Assumptions*

In contemporary science theories are expressed by systems of symbols, or symbolisms. What these symbols stand for is more or less clear from (*a*) the formulas in which they occur and (*b*) the explicit designation rules that assign constructs to those symbols. An example of a *designation rule* (or proposal) is: ⌜Let '*S*' designate (name, represent, stand for, symbolize) a set⌝. Such rules are conventional in the sense that the precise symbol chosen to represent a construct is immaterial as long as every time the symbol occurs it be assigned the same construct. The resulting mathematical formalism is not an abstract mathematical theory (e.g. the general theory of groups) but an interpreted one – e.g. a theory describing a group of transformations of the Euclidean plane.

Such a mathematical *formalism* is by itself neutral with respect to matters of fact. So, unless the formalism is "read" in factual terms, it will "say" nothing about reality. Take, for example, the mathematical theory of natural selection – the hard core of the contemporary theory of evolution. As Waddington (1967, p. 14) says in colorful terms, "the whole real guts of evolution – which is, how do you come to have horses and tigers, and things – is outside the mathematical theory (...) The sheer mathematical statement is largely vacuous. The actual way it is applied, not by the mathematical theorist but by the biologist working with the subject, is not vacuous at all". This does not mean that "the whole real guts" of a scientific theory has to remain detached from its formalism: it can and must be blended with it. That is, the mathematical formalism becomes a factual theory, or rather one of a number of possible factual theories with the same underlying formalism, if the suitable factual interpretation is supplied.

The typical situation is exhibited in the following block diagram. (For a detailed analysis see Ch. 6.)

Abstract theory	Mathematical interpretation →	Mathematical formalism	Factual interpretation →	Factual theory

The factual interpretation of a theory is superimposed on a mathematical framework with a definite mathematical meaning and is determined by two disjoint sets of semantic rules. One is formed by the *denotation rules* or symbol-thing correspondences identifying the referents of the theory. This set constitutes what Campbell (1920, pp. 122–128) called the "dictionary" of the theory. The other is the set of *semantic assumptions* or function-property correspondences. Whereas the former point to and baptize the referents of the theory, the semantic assumptions link constructs to factual items by indicating the traits of things that the constructs are supposed, rightly or wrongly, to represent. A few examples will help getting these ideas across.

Example 1 In mechanics we find, among others, the following semantic formulas:

*DR*1 π *denotes* (names, stands for) a particle.

*SA*1 $X(\pi, f, t)$ *represents* (or measures) the position of the particle π relative to the reference frame f at the instant of time t.

Example 2 In the genetics of populations we meet

*DR*2 Let *a be* (denote, name, stand for) an allele.

$SA2$ *W dt represents* the total increase in fitness of the population of interest through the time interval *dt*.

Example 3 In mathematical sociology we come across

$DR3$ Let k *denote* a rung in a hierarchy.
$SA3$ $\sum_k k \cdot n_k$ *represents* the status of an individual with n_k subordinates on the kth rung of the hierarchy.

Because names are conventional, denotation rules are partly conventional. That is, the same factual items might be relabelled without qualms, as this would ensue in a different symbolism not in a new body of theory. However, inasmuch as the denotation rules indicate the hypothetical referents of the theory, they are not conventional: the theory might turn out to refer to different entities or to no real entities at all. And it is even clearer that the semantic assumptions are full blown hypotheses not just a matter of notation. They are not hypotheses about reality but about the theory-reality correspondence. Consequently a change in the semantic assumptions of a theory will ensue in a different theory with the same mathematical formalism. For example $SA1$ above is unacceptable to an operationist, who would reformulate it in terms of the measurable position values obtained by an observer attached to (or constituting) the frame f. As to $SA2$, it might not hold in a theory other than Fisher's mathematical theory of natural selection. Finally $SA3$ might read differently in an organization theory focusing on effective power instead of status.

The distinction we have drawn is seldom made explicitly in the scientific literature. Here one finds careless statements such as ⌜T is the absolute temperature⌝, which may be interpreted either as a designation rule or as a semantic assumption. It behooves the semanticist and the foundations worker to find out in each case whether the blanket word 'is' is intended to stand for "designates", "denotes", or "represents". Such distinctions are not pedantic: they make the difference between convention and conjecture. And this difference is of course of paramount methodological difference. Yet it is not made in the contemporary philosophy of science, where semantic rules and semantic hypotheses go by the common name of *correspondence rules.*

Since denotation rules are only partly conventional and semantic assumptions are wholly hypothetical, they ought not to be accepted on authority as they sometimes are. It should be possible to argue about them and even to put them to the test – though not independently of the mathematical formalism they serve. Let us try and explain. A theory constitutes no factual theory unless it includes a set of semantic formulas interpreting its basic concepts in factual terms. (Whether these semantic ingredients are displayed within the body of the theory, or hinted at in casual remarks, they are attached to the mathematical formalism.) Consequently it is the theory as a whole, i.e. the mathematical formalism together with the set of semantic formulas, that is subject to empirical tests. Fancy sending either a fleshless formalism or a boneless semantics to the laboratory.

If the empirical tests prove favorable and are themselves reliable, the formalism-semantics composite is declared confirmed until new notice. Otherwise there are three possibilities: to blame the formalism, to blame the semantic formulas, or to scrap both. The former alternative invites mending the formulas without touching the interpretation: this is a common enough procedure and one that succeeds often. The third alternative calls for a fresh start and may end up in a scientific revolution such as the founding of quantum mechanics. The middle alternative, viz., repairing the semantic formulas, seems to be less well known to philosophers but it is tried more often than radical upheavals. The following examples are well known to physicists and they show how significant a change in the semantics of a theory can be.

Example 1 The main impact of special relativity upon classical electrodynamics was forcing it to abandon all reference to the mechanical ether. From then on the theory was deemed to refer to, and represent, electromagnetic fields. Consequently all the questions concerning the properties of the ether and motions relative to it disappeared overnight. *Example 2* In the beginnings of wave mechanics the ψ function was regarded either as a real wave or as a mere mathematical auxiliary (an intervening variable). Later on its square was assumed to represent the mass density of the system associated with ψ. Eventually the so called statistical interpretation was adopted on reasonable grounds and all the rival interpretations shown to be responsible for consequences incompatible with the empirical evidence. *Example 3* Yukawa's theory of nuclear forces assumed that it was concerned with μ-mesons. But the latter refused to conform to the theory. Eventually it was found that π-mesons did satisfy the theory reasonably well. Accordingly the original semantic assumption was changed.

The semanticist cannot decide whether to change the formalism or the semantics of a scientific theory in the face of adverse empirical evidence. All he can do is to insist that the semantic formulas be formulated explicitly and clearly in order to better keep them under control. He can also alert to the philosophical concomitants of any given semantic – but this point deserves another section.

4.2. *Philosophical Commitment of the SA's*

The foregoing examples of semantic formulas (denotation rules and semantic assumptions) suggest laying down the following methodological

RULE In a well formulated scientific theory

(i) the referents should be indicated by explicit *denotation rules* bearing on the class symbols of the theory;

(ii) a denotation rule has the form: 'σ denotes a (member of the class or species) Σ';

(iii) explicit *semantic assumptions* should establish which if any property of the referent(s) does a basic predicate of a theory represent;

(iv) a semantic assumption has the form: 'P represents the... of a σ in Σ', where the blank names the property or relationship represented by the predicate P;

(v) the denotation of every defined symbol, as well as the representing function of every defined concept, should be consistent with the semantic formulas in which the defining items occur.

Needless to say, reference and property assignments must be consistent with the structure of the symbol or construct concerned. Thus if a single class symbol occurs in a theory then it can denote a single kind of entities rather than, say, the set of ordered pairs thing-apparatus or thing-observer. (This condition is ignored by the operationist interpretations.) And if a certain function is not time dependent then it cannot represent a change in the course of time. Secondly: it is not mandatory for every predicate in a theory to represent some property. Many predicates, particularly in sophisticated theories, represent no definite property even when (*a*) they have definite referents and (*b*) they help defining constructs that do represent. This is the case with lagrangians, partition functions, and the eigenfunctions of quantum mechanical operators other than the energy operator. Thirdly, denotation rules and semantic assumptions *sketch* factual meanings but do not exhaust them. No single piece of a factual theory, not even its semantic assumptions, gives a full characterization of the factual meaning of the theory: only the theory as a whole is fully meaningful. More on this in Ch. 5, Sec. 6.

The preceding rule can be implemented in alternative ways for any given theory. In other words, given a mathematical formalism and the above rule, there is still play enough to obtain a number of alternative factual theories – as many as alternative sets of semantic formulas be superimposed on the formalism. Which is just as well, for we want to try out different theories before we choose one of them.

Moreover, the implementation of the preceding rule depends upon one's philosophy. In particular, a semantic assumption may be construed in at least three different ways:

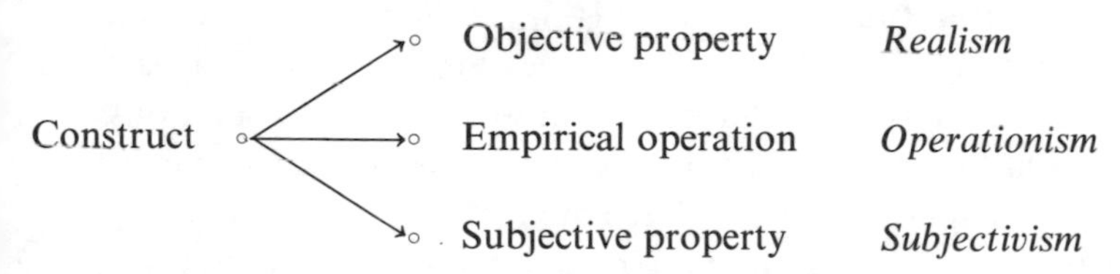

In the first case the theory, even should it refer to brains, would be assigned a strictly objective meaning, i.e. one independent of the cognitive subject or observer. In the second case the theory, even should it refer to galaxies far out of reach, would be assigned an operational meaning, i.e. every predicate in it would be correlated to an object-subject compositum. And in the third case the theory would concern the theoretician himself, e.g. his state of knowledge, the strength of his beliefs, and the span of his uncertainties.

These three lines in the semantic spectrum are particularly bright in relation with stochastic theories. Thus a transition probability $P(a\rightarrow b)$ from state a to state b of some system is usually interpreted in either of the following ways.

Realist SA $P(a\rightarrow b) \hat{=}$ The *tendency* for the system to evolve from state a to state b.

Operationist SA $P(a\rightarrow b) =$ The *frequency* with which the system, when subject to some experimental condition, is *observed* to evolve from state a to state b.

Subjectivist SA $P(a \to b) \triangleq$ My rational degree of *belief* in the system's transition from state a into state b.

Often either of these semantic assumptions is adopted on faith or on the strength (or weakness) of some philosophical tradition, and in any case with little concern for the structure of the concept concerned or for the role it plays in the law statements of the theory. If the probability function P occurring above is analyzed it is found that a and b denote states of some system quite apart from any experimental considerations: if there is a single object variable involved then there is no room for a second object variable such as an apparatus, much less for a third object variable representing an observer. The object variable must then represent either an external object or the theoretician (or his mind). If the former, then all the law statements in which P occurs must concern external objects whence they must be testable by handling and observing such objects. But if the referent of P is the theoretician himself, then all the statements in which P occurs must likewise refer to him (at least) and so the empirical tests of such formulas must include introspections.

We should then realize what our semantic assumptions commit us to and what their philosophical underpinnings are. And we must demand that every semantic assumption, far from being accepted on authority, be *testable* both conceptually and empirically, namely thus:

(i) A semantic assumption should *fit the structure* of the concept concerned and it should not violate any of the basic formulas in which the construct occurs. Checking this condition is of course a matter of pencil and paper: it is a conceptual test.

(ii) The construct involved in a semantic assumption should in fact describe what it is assumed to represent. Checking this condition calls for *empirical tests* of some of the formulas interpreted by the semantic assumptions.

Thus if a theory contains semantic assumptions of the operationist type, we should check (*a*) whether the mathematical formalism does make room for such assumptions (e.g. whether there are enough variables to represent not only the system but also the experimental set up and the observer), (*b*) whether the theory allows one to compute quantities representing properties of something that is not being subjected to experimental conditions, and (*c*) whether the experimental results are in fact critically dependent upon the observer and his equipment. And if the theory contains semantic assumptions of the subjectivist type, we should check (*a*) whether the theory does issue predictions about the theoretician's (or the observer's) mental states or behavior and (*b*) whether the available empirical data do bear on the cognitive subject himself rather than on objects external to him – e.g. other people. In conclusion, the semantic assumptions of a scientific theory are not conventions and they are not *dicta* beyond controversy: they are testable hypotheses. (Only, they cannot be put to the test separately from the formulas they endow a factual content with.) That they are hypotheses, and often controversial ones, can be seen from the debates on the interpretation of the mathematical formalisms of the quantum theories, to which we now turn.

4.3. *Application to Quantum Mechanics*

Quantum mechanics is probably the scientific field with the largest number and variety of semantic assumptions. (See Bunge (1956) and 'Quantum-mechanics debate', *Physics Today* **24** (1971), No. 4, pp. 36–44.) Quite often pairs of mutually incompatible semantic assumptions are nonchalantly attached to a given formula in one and the same paper. Table 3.5 exhibits a modest sample of such alternative semantic assumptions in relation to just two formulas of daily use. These formulas are

the *eigenvalue equation* $A_{op}u_k = a_k u_k$, where $A_{op} \hat{=}$ Property A,

and

the *eigenfunction expansion* $\psi = \sum_k c_k u_k$.

Only a detailed analysis of the whole theory enables one to side with one or the other alternative set of semantic assumptions. An analysis performed elsewhere (Bunge, 1967b, 1967e, 1969) shows that only the realist interpretation is allowed by the mathematical formalism of quantum mechanics. We cannot go into details here but we will give a couple of reasons for rejecting the standard or operationist (or Copenhagen) version of the semantics of quantum mechanics.

A first reason why a semantic assumption cannot correlate theoretical constructs with empirical items is that the former are just not roomy enough to make place to instruments and observers. In other words, the

TABLE 5

Two sets of rival semantic assumptions (operationist and realist) for the eigenvalue equation and the eigenfunction expansion. A_{op}: operator other than hamiltonian

Symbol	Mathematical status	Operationist semantic assumption	Realist semantic assumption
A_{op}	Operator in a Hilbert space	Refers to an object-apparatus-observer block. Represents a dynamical variable of the whole.	Refers to a physical system irrespective of environment. Represents intrinsic property of system.
u_k	Function in a Hilbert space	Same referent as A_{op}. Represents state of block when measurement yields value a_k.	Refers to system (no environment). Represents nothing. (Eigenfunction expansions are mathematical auxiliaries.)
a_k	Real number	Possible value one finds upon measuring dynamical variable represented by A_{op}.	Possible value of dynamical variable A_{op}.
ψ	Function in a Hilbert space	State of observer's knowledge before performing any measurement.	State of physical system in some environment (eventually null or nonexistent).
$\|c_k\|^2$	Positive real number	Probability of getting value a_k upon measuring dynamical variable represented by A_{op}, or strength of one's belief that a_k will turn up upon measurement.	Propensity or tendency of value a_k of A

formulas do not contain enough variables that could refer to experimental situations: this is a semantic argument. A methodological reason why a semantic assumption cannot assign empirical items to a theoretical construct is that theories cannot possibly concern their own tests. Every empirical test ropes in objects other than those represented by the theory, namely pieces of apparatus. And these additional objects must be represented by additional theories. These, the auxiliary theories representing the object-experimental set-up complex, will differ according to the nature of the experimental set-up, which is rarely unique. For example, thermodynamics contains the pressure concept but no provisions for designing pressure gauges. And the latter are often employed in testing quantum mechanics but the very notion of pressure is absent from the latter. Another example: electromagnetic fields affect the growth of plants and their shape. Hence one might think of using plants as low sensitivity instruments for measuring some traits of the electromagnetic field. But it would be preposterous to claim that the electromagnetic theory, let alone quantum electrodynamics, represent plants. In general, any statements concerning the nature or the strength of the empirical evidence relevant to a scientific theory are built with the help of at least one other theory (Bunge, 1967a, 1973b). However, this is another matter – one for methodology not semantics.

5. Conclusion

In this chapter we have initiated an investigation of the notion of representation which will be completed in Ch. 6, Sec. 3. This notion is conspicuously absent from the conventionalist, formalist, and empiricist semantics of science. In particular the latter, which is the best known, has no use for the concept of representation because it rejects the realist thesis and holds nonlogical concepts to be ultimately percepts or logical constructions out of the latter. According to empiricism scientific theories have an observational or empirical content given them by the "correspondence rules" of the theory, which rules link the theoretical terms to observational terms, which are in turn theory-free. The notion of a real or autonomous thing, which to the realist is the object of both theory and scientific experience, does not occur at all in that doctrine.

An analysis of real specimens of scientific theory would have shown that the sole support enjoyed by empiricist semantics is tradition – a tradition of exact analyses and careful reconstructions of a nonexistent science. A case study work of the type sponsored by Törnebohm's Institutionen för Vetenskapteori in Göteborg would have shown that (*a*) the aim and outcome of theorizing is not canning sensations or even data but representing selected aspects of putatively real things, (*b*) theoretical concepts are sophisticated mathematical constructions that can neither be defined in terms of empirical operations nor construed as logical functions of observational concepts, and (*c*) empirical tests, which are conducted in order to estimate the truth value of hypotheses and theories, consist in operations planned in the light of further theories – a process that Agassi has aptly named a *bootstrap* operation.

Experience – controllable not subjective, refined not coarse – is the methodological link between theory and reality. This link does not belong to the theory: if it did then either of them would be redundant. And it is not the sole bridge that spans the chasm between theory and reality: there is also the semantic bridge constituted by the semantic assumptions of the theory. (We do not include among them the semantic conventions or designation rules, which are conventional symbol-construct links.) Those hypotheses are of two kinds:

(a) *denotation rules* of the form "Symbol s denotes thing θ";

(b) *representation assumptions* of the form "Function F represents property P of thing θ".

The semantic assumptions of a scientific theory couple symbols and their designata with supposedly real things and their properties. Since they have putatively real referents and fail to be tautologous, by Ch. 2, Sec. 4 our semantic assumptions are factual statements. Hence they have a factual content. Moreover since they are the only systematic indication concerning whatever the theory is supposed to represent, the semantic assumptions determine, at least in outline, the factual meaning of the extralogical constructs involved in them. In other words, the corrigible assumptions concerning what concepts in a theory represent which features of the world contribute to endowing a scientific theory with a factual content. Which takes us to the subject of the next chapter.

CHAPTER 4

INTENSION

⌜People make mistakes⌝ and ⌜People learn to correct some mistakes⌝ have the same referents but different senses. They represent different traits of mankind, and this quite apart from their truth values. Yet the senses of these two statements, whatever 'sense' be taken to signify, cannot be disjoint since the second proposition implies the first. The question is: What exactly are their senses and how are they related? This is the problem we face in this chapter. However, a complete solution to it will ensue only as a merger of the present chapter with the next: in this chapter only one sense of "sense" will be investigated.

1. FORM IS NOT EVERYTHING

1.1. *Concepts of Sense*

Whereas the concept of form is presumably clear, that of content is reputedly obscure. However, there is no denying that we handle contents all the time. Thus we are likely to praise logic (or to debase it, as the case may be) for being content free, hence for treating ⌜Dolphins are mammals⌝ on a par with ⌜Fractions are real numbers⌝. But it is equally clear that in order to find out the form of a construct we must know something about its content or lack of it. Change the content and the form may alter. For example, *prima facie* "longer than" is a dyadic relation but, when analyzed in relativistic physics, it proved to be triadic: here we must write ⌜x is longer than y relative to z⌝.

Scientific constructs come with both a form and a content. These two aspects can be distinguished but not separated: only logic can afford to deal with pure form. But distinguished they must be if we wish to understand what makes logic different from the rest, and if we are to explain the recurrence of the same mathematical structures in different fields. Take for instance "fitter than" and "more intelligent than". The two are isomorphic in the sense that they are different interpretations of the relation $>$, and so they have different contents. Even coextensive concepts, i.e. concepts with the same instances, may have different contents. For example, "mammal" and "hairy" are isomorphic and coextensive, yet they have ostensibly different contents. Likewise "avian" and "feathery", "the successor of 1" and "the smallest prime number", and so on and so forth. In short, a scientific construct is not characterized by its form and extension alone.

Granted then that, outside logic, there is something, variously called 'content', 'sense', 'intension', or 'meaning', that is not to be written off. Granted also that reference and extension are of little if any help in determining contents, as coreferentials and even coextensives, such as "human" and "cruel", can have different senses. Granted, in sum, that contents or senses are *sui generis* objects, distinct from forms, reference classes, and extensions, and thus far more difficult to pin down than either of these. The question is: What do we do with those elusive objects? Two attitudes are possible: retreat or attack. The former is argued for in this way: "Senses have always been obscure. They have defied the best philosophical minds. Therefore they are hopelessly obscure. Consequently we had better give up any attempt to clarify them". The outcome of this policy is to allow the beast to roam in the wilderness of obscurantist philosophy: after all, "man is a sense-making animal" (Quine, 1966, p. 175), and nothing human is reputedly alien to philosophers. We refuse to adopt the defeatist stand. We propose to launch an attack with a view to taming the wild beast. Our motto will be "Divide and conquer".

In fact our point of departure is recognizing that there is not a single concept of sense but rather three different concepts. We shall show later on that they are related, but we begin by distinguishing them as so many dimensions of sense. Indeed we shall propose and explain the thesis that a construct may have the following kinds of sense:

(i) the totality of its conceptual determiners, or *purport*:
(ii) the set of constructs it subsumes or embraces – its *intension*;
(iii) the totality of its *implicates*, or *import*.

In each case the sense of a construct is a set of constructs. (Note the difference with respect to both reference and extension: a scientific construct will refer to, and extend over, sets of factual items.) In each case the content of a construct is logically related to the construct itself. (Note again the difference with respect to both reference and extension.) But in the case of purport we look upwards, at the impliers; in the case of intension we look horizontally, into the construct; and in the case of import we look downwards, at the consequents:

Upward view — Determiners (*purport*) of c

Horizontal view — What c embraces or subsumes: the *intension* of c

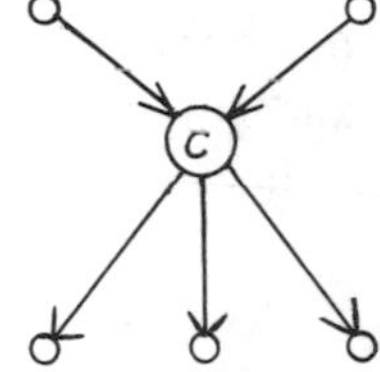

Downward view — Implications (*import*) of c

Example The concept Δ of Euclidean triangle.

Purport (Δ) = The (finite) set T of statements in Euclidean geometry sufficient to characterize all the properties of Δ.
Intension (Δ) = {Δ is a closed plane figure with 3 rectilinear finite sides}.
Import (Δ) = {Δ is a polygon, The inner angles of Δ add up to π, etc., etc.}
Referents (Δ) = The (infinite) set of plane figures.
Extension (Δ) = The (infinite) subset of plane figures satisfying the conditions T of triangularity.

Of these three concepts of sense that of purport seems to be new although it sounds quite natural. It will be shown to embrace the concept of gist, or basic purport, of everyday use. The second concept of sense, viz., "intension", seems to elucidate the *significatio* of Medieval logicians as well as the *compréhension* of the Port Royal *Logique* (Arnauld and Nicole, 1662). The latter was translated by Sir William Hamilton as *intension*, a term that has been with us – and often against us – ever since (see Kneale and Kneale, 1962). A similar concept occurs in Bolzano (1837), Meinong and many others as *Inhalt*. Frege used it under the label of *Sinn* (sense) though without subjecting it to an exact theory. And, at least formally, our concept of intension is close to the *L-content* of Carnap (1942). Finally the third dimension of sense, namely import, is near the *C-content* of Carnap (1942), the *intension* of Lewis (1944, 1951) and of Castonguay (1971), and the *truth content* of Popper (1963b, 1966).

Our three concepts of sense have then roots in common sense or in technical philosophy. However, our theories of them exhibit important differences with respect to previous theories. For example, our theory of import does not involve any notion of truth: it is a purely syntactic theory employing the algebraic theory of filters. Besides, our calculus of import joins with our views on the nature of a factual theory, to yield new ideas on factual content as well as on meaning changes. Moreover we shall show that the three concepts – those of purport, intension, and import – are interrelated. (For example, purport and import are mutually dual in the algebraic sense, and in the case of axiomatic systems the intension is included in the union of purport and import.) Finally it will be seen that none of our concepts of sense can be equated with that of semantic information, and that they must also be kept separate from the concepts of testability.

Since we shall deal systematically with three different concepts, we may expect one theory for each. In point of fact we shall evolve the three theories independently and shall explore their relations at the end. We shall find it convenient to start at the middle, i.e. from intension. Purport and import will be tackled in the next chapter. But before proceeding to build a theory of intensions we must show why we need one and we must guard against confusing 'intensional' with 'intentional', 'nonextensional', and 'non truth functional'.

1.2. *Extension Insufficient*

The set theoretic and model theoretic treatments of mathematical concepts have been so effective, and they have had such a clarifying and unifying power, that they have buttressed the belief that only extensions (domains of individuals) matter. This belief may be spelled out in the following tenets:

E1 Every mathematical entity is either an individual or a set.

E2 Reference is the same as extension.

E3 Senses or intensions are either ghostly or reducible to extensions.

Criticism of E1 This thesis is particularly apparent in the theory of relations as treated in set theory. Here it takes the form: "Every relation may be defined simply as a set of ordered n-tuples". This claim is exaggerated: the predication relation in logic (i.e. the relation between a predicate and its subject or subjects, as in $\ulcorner Pa \urcorner$), the membership relation in set theory, and the relation of satisfaction in model theory, are not definable as sets of ordered couples. Hence although sets, in particular the extensions of relations, are extensional objects, their theories are not purely extensional. What is true is that, in *most* cases in *pure* mathematics (not so in factual science), only the extension of a relation *need* be taken into consideration for purposes of *proof*. In any case it is incorrect to identify a relation with its graph or extension. Bourbaki, for one (or rather many), draws a careful distinction between a relation and its graph (extension), in particular between a correspondence (e.g. a function) and its graph. A correspondence between the sets A and B is defined as a certain triple $F = \langle G, A, B \rangle$, where G is the graph of F (Bourbaki, 1970, Sec. 3). Finally, it is not true that all mathematical theories are reducible to set theory – nor, *a fortiori*, that they can dispense with intensions. The basic concepts of category theory, namely those of object and morphism, are not so reducible. It is rather the other way around: the set theoretic predicates may be defined in categorial terms (Lawvere, 1966). In sum, the extensionalist program of reducing every mathematical object to either an individual or an aggregate of individuals subject to the postulates of (some) set theory, though remarkably successful, is not fully feasible.

Criticism of E2 To begin with, while both predicates and propositions can be assigned referents, only predicates are normally said to have an extension. And even in the case of predicates a distinction between extension and reference class is needed. For example,

$$\mathcal{R}\,(\text{Prime number}) = \text{Natural numbers},$$
$$\mathcal{E}\,(\text{Prime number}) = \text{Prime numbers}.$$

The difference is even more pronounced outside mathematics. Thus

$$\mathcal{R}\,(\text{Flying on a broomstick}) = \text{Witches},$$
$$\mathcal{E}\,(\text{Flying on a broomstick}) = \emptyset.$$

A more respectable example is this:

$$\mathcal{R}\,(\text{Rational}) = \text{Mankind}, \ \mathcal{E}\,(\text{Rational}) \subset \text{Mankind}.$$

Finally an example from recent science: while the (hypothetical) reference class of "parton" is the set of partons, nobody knows what the actual extension of that theoretical predicate might be, not even whether it is nonvoid. In conclusion, *Reference* $\neq$ *Extension*. And, as we shall see in Ch. 9, Sec. 1.1, in general extensions are not subsets of reference classes: only the extensions of unary predicates are.

Criticism of E3 The thesis that senses can be written off altogether is false even for set theory, the paradigm of a "purely extensional theory". Indeed set theory distinguishes a predicate from its extension. So much so that it devotes an entire axiom (schema) to the relation between a predicate and its extension, namely the principle of abstraction or its improved version the principle of separation (*Aussonderung*). According to this axiom schema, to every well formed formula Px of the lower predicate calculus and to every set A there corresponds another set B whose elements are exactly those members x of A for which Px holds (cf. Suppes, 1960, p. 21):

$$(\exists B)\,(x)\,(x \in B \Leftrightarrow x \in A \;\&\; Px).$$

In our terminology A is included in the *reference class* of P while $B = \mathscr{E}(P)$ is the *extension* of P. In the case of monadic predicates, like the one involved in the previous formula, $\mathscr{E}(P) \subseteq \mathscr{R}(P)$. *Example.* Let $A = \mathbb{Z}$ = The set of integers, and $Px = \ulcorner x$ is prime$\urcorner$, with $x \in \mathbb{Z}$. Then $\mathscr{R}(P) = \mathbb{Z}$ and $\mathscr{E}(P)$ = The set of prime numbers $\subset \mathbb{Z}$. The axiom of separation is not a device for getting rid of whatever sense a predicate may have: it just clarifies the *relation* between a predicate (whichever its sense may be) and its extension. Finally, the extensionalist program fails altogether in factual science. For example, the predicates "metallic", "good thermal conductor", and "good electric conductor" have the same extension although their senses are obviously different. If only for this reason the predicates occurring in theoretical science cannot be defined *within* set theory, i.e. in the last analysis solely in terms of membership in some set, as Suppes (1967, 1969) has been advocating untiringly. We also need semantic assumptions concerning reference and representation, as we have seen in Chapters 2 and 3.

To sum up, the extensionalist thesis is false even within mathematics: we may refrain from looking at intensions for a while but we cannot suppress them. Nor should we mistake them for other characters, such as psychological intentions and modalities. This confusion deserves a separate subsection.

1.3. *'Intensional': Neither Pragmatic nor Modal*

Principia Mathematica (Whitehead and Russell, 1927, pp. 72ff and 659ff) misled several generations of philosophers by its quaint use of the word 'intensional'. What *PM* meant by this term was not one of the *semantic* concepts of sense but a certain class of *pragmatic* concepts, often called 'propositional attitudes', "such as what somebody believes or affirms, or the emotions aroused by some fact". Typical "intensional" terms would be 'doubts', 'is puzzled by', 'believes', 'knows', and 'asserts', which designate relations between a person and a proposition. Correspondingly a statement of the form $\ulcorner x$ believes that $y\urcorner$ is often called an *intensional context*. And any attempt to clarify and systematize such pragmatic statements is often said to belong to *intensional logic*.

To find out what such misnomers designate let us examine a typical example of an "intensional context". Consider the statements

$$p = \ulcorner \text{Milk is good} \urcorner \tag{1}$$

and

$$q = \ulcorner \text{Baby believes that } p \urcorner. \tag{2}$$

Whereas p is allegedly a perfectly "extensional" construct, q would be an "intensional" one, for its truth value is not a function of the truth value of the subordinate statement p alone. If it were, the truth value of q would remain invariant upon replacing p by any of its equivalents – which is not the case. Indeed, p is equivalent to

$$r = p \,\&\, (s \vee \neg s) \tag{3}$$

which, substituted for p in (2), yields

$$t = \ulcorner \text{Baby believes that milk is good and } (s \vee \neg s). \urcorner \tag{4}$$

which, unlike q, is false.

All this has nothing to do with intensions although it might be claimed to concern intentions. In fact, since tautologies are void of content, p and r above have the same intension. But they fail to have the same effect on all persons: whereas to a logician the utterance of r conveys just as much information as p does, to a layman r may sound more informative than p and to a baby r is likely to sound as gibberish. (More on information in Sec. 3.2.) All of this is pragmatics not semantics.

The actual situation is simply this. The statement q in (2) depends not just on the statement p but also upon a person, namely Baby, and a circumstance or time t. In brief, $q = B(\text{Baby}, p, t)$ where 'B' stands for the belief function. In general, pragmatic statements involving propositions are of the form: $q = P(\text{person}, p, t)$, where '$P$' stands for a pragmatic function such as that of knowing. No wonder logic cannot cope with such hybrids. Little wonder that the truth value of such a statement q depends not just on the truth values of its subordinate statement p but also on the background knowledge and the state of the person concerned. In other words P is not a truth function – more accurately, it is not a truth preserving map. P is a *non truth preserving pragmatic function* – or, if preferred, an inten*t*ional function. To call it 'inten*s*ional' is to equivocate.

Pragmatic functions constitute a subclass of the class of non truth preserving functions. Some of the latter may not involve any person in an overt way. For example, ⌜p is conceivable⌝, ⌜p is doubtful⌝, ⌜p is puzzling⌝, ⌜p is beautiful⌝, ⌜p is suggestive⌝, ⌜p is useful⌝, and ⌜p is provable⌝ are not ostensibly pragmatic statements. Yet they can be construed as covert pragmatic statements. For example, ⌜p is conceivable⌝ may be construed as short for ⌜There is at least one person who can conceive of p⌝. But this is irrelevant to semantics. What is relevant to it is what all such statements have in common. And this is that they resist the substitution of equivalents. More explicitly: In all of these cases the subordinate proposition may not be replaced by any of its logical equivalents *salva veritate*, i.e. with preservation of the truth value of the total proposition. To paraphrase *PM*: A function of a construct will be called a *truth preserving function* when its truth value with any argument is the same as with any formally equivalent argument. Otherwise it will be called a *non truth preserving function* – by no means an "intensional function".

Modal statements are of course the best studied of all those that violate the principle of the substitutivity of equivalents. Since they, too, fall outside ordinary ("extensional") logic, Lewis and Carnap concluded (probably misled by the unorthodox use of 'intensional' in *PM*) that they must obey some systems of "intensional logic". Thus we have a first confusion: "Whatever is modal is intensional". Now indulge in the most common fallacy and obtain "Whatever is intensional is modal". These two mistakes engender the now rather popular misconception that modal logic is the key to whatever boxes ordinary logic cannot open: not only possibility but also intensions and "propositional attitudes". This conflation of "modal" with "intensional" can only vitiate semantics.

We shall keep clear of these confusions and will make no use of that all purpose tool, modal logic. The very idea of an "intensional concept", as opposed to an extensional one, is a misfit: if a concept has an intension it also has an extension even if nil. (The converse is false: sets have no intensions.) Hence the expressions 'intensional function' and 'intensional context' are misnomers. Frege's name for them, namely *ungerade* (oblique), and Quine's 'referentially opaque', are preferable. However, since 'referential opacity' has at least three different significations (Sharvy, 1972), we prefer the longer but nonmetaphorical and more explicit name 'non truth preserving'. And we shall gladly leave the study of non truth preserving functions (or inten*t*ional concepts) to pragmatics, epistemic logic, modal logic, and any other discipline that be deemed competent to handle them.

We now turn to a study of intensions conceived as semantic objects.

2. A CALCULUS OF INTENSIONS

2.1. *Desiderata*

Our aim is to build a calculus of intensions enabling us (*a*) to exactify the notion of intension and (*b*) to compute the intension of a complex construct, such as a conjunction, out of the intensions of its components. And we wish our calculus to formalize and articulate the following intuitive desiderata:

D1 All and only predicates and statements shall have an intension.

D2 The intension of a construct shall be a set.

D3 The intension of a construct in a context $C=\langle S, \mathbb{P}, D\rangle$ shall be comprised between $\emptyset$ and C, i.e. if $P\in C$, then $\emptyset\subseteq\mathcal{I}(P)\subseteq C$.

D4 If P and Q are both either predicates or statements, then

(*a*) $\mathcal{I}(P \,\&\, Q)\supseteq\mathcal{I}(P), \mathcal{I}(Q)$

(*b*) $\mathcal{I}(P\vee Q)\subseteq\mathcal{I}(P), \mathcal{I}(Q)$

D5 If $\mathcal{I}(P)\subseteq\mathcal{I}(Q)$, then

(*a*) $\mathcal{I}(P \,\&\, Q)=\mathcal{I}(Q)$

(*b*) $\mathcal{I}(P\vee Q)=\mathcal{I}(P)$

D6 If two constructs are identical so are their intensions.

D7 Tautologies are intensionally minimal and contradictions are intensionally maximal.

D8 The intension of $\ulcorner P\&(P\Rightarrow Q)\urcorner$ is the same as that of $\ulcorner P\&Q\urcorner$.

D9 Cointensives are coextensive – but not conversely.

D10 The smaller the intension of a construct the larger its extension.

We proceed to exhibit an axiom system that satisfies all these desiderata. (For interesting alternatives see Leonard (1967), Suszko (1967), Castonguay (1972), and Weingartner (1973).)

2.2. *Principles and Definitions*

Intension shall be construed as a certain function that assigns to every element of a conceptual universe of discourse a certain subset of the latter. More precisely, we lay down the following axiomatic

DEFINITION 4.1 Let U be a set of predicates or of statements. The *intension* function $\mathscr{I}$ is the function from U onto the family $\mathscr{P}(U)$ of subsets of U such that, for any P and Q in U,

(i) If $P\ \&\ Q$ is defined, then $\mathscr{I}(P\ \&\ Q)=\mathscr{I}(P)\cup\mathscr{I}(Q)$;

(ii) $\mathscr{I}(\neg P)=\overline{\mathscr{I}(P)}$;

(iii) If $P=Q$ then $\mathscr{I}(P)=\mathscr{I}(Q)$.

Example 1 $\mathscr{I}$(Female human)$=\mathscr{I}$(Female)$\cup\mathscr{I}$(Human). *Example 2* $\mathscr{I}$(Unattached)$=\overline{\mathscr{I}(\text{Attached})}$. *Example 3* Since $\ulcorner 1+2=3\urcorner$ is the same proposition (though not the same sentence) as $\ulcorner 3=1+2\urcorner$, $\mathscr{I}(1+2=3)=\mathscr{I}(3=1+2)$. Note that, pragmatically, the two sentences are different: while the first expresses the unique outcome of a correct addition operation, the second summarizes one of the possible results of a correct decomposition of 3.

Remark 1 The generalization of the first axiom to multiple conjunctions, whether finite or infinite, is immediate:

$$\text{If } \bigwedge_i P_i \text{ is defined, then } \mathscr{I}(\bigwedge_i P_i)=\bigcup_i \mathscr{I}(P_i) \text{ for } P_i\in U,\ i\in N.$$

Remark 2 Definition 1 tells us, so to say, how intensions behave not what they are. Indeed, it fails to prescribe how to find the intensions of the constituents of the compound constructs. (Likewise truth functional logic shows us to compute the truth value of a compound statement given the truth values of its components.) The problem of determining the basic intensions will be faced in the next Chapter. *Remark 3* The converse of clause (iii) in Definition 1 is false. Thus ⌜Temperature $(b)=t$⌝ and ⌜Temperature $(b')=t'$⌝, though cointensive, are not identical (unless of course the referents b and b' happen to be the same).

We can now elucidate certain interesting derivative concepts.

DEFINITION 4.2 A construct is said to be *intensionally empty* iff its intension is equal to the empty set.

DEFINITION 4.3 A construct in U is said to be *intensionally universal in U* iff its intension equals U.

DEFINITION 4.4 If P and Q are in U, then P is said to be *intensionally included* in Q (or intensionally poorer than Q) iff $\mathcal{I}(P) \subseteq \mathcal{I}(Q)$.

DEFINITION 4.5 Two constructs in U are said to be *cointensive in U* iff their intensions are the same: $\mathcal{I}(P) = \mathcal{I}(Q)$.

DEFINITION 4.6 Two constructs are said to be *intensionally independent* ("perpendicular") iff their intensions are disjoint:

$$P \perp Q =_{df} \mathcal{I}(P) \cap \mathcal{I}(Q) = \emptyset.$$

DEFINITION 4.7 Two constructs are called *intensionally dependent* iff they are not independent – i.e. if their intensions overlap.

Before looking at the consequences of our assumptions and definitions let us note that an intensionally empty construct is a construct all right – not a nonconstruct. For example tautologies will prove to be intensionally void (Theorem 5). The intension function $\mathcal{I}$ is defined for predicates and statements and for them only, hence sets and nonconstructs fail to have a definite intension. One such nonconstruct results from the conjunction of predicates defined on disjoint domains, such as "wind" and "stupid". Our axiom (i) does not apply to "stupid wind" (Yeats) because there is no nonvoid domain on which "stupid wind", or any other such freak, is defined. In other words, for a predicate to have a definite intension, even a null one, it must be well constructed to begin with:

recall the rule in Ch. 1, Sec. 1.3. This does not entail throwing poetry away but just noting that some poetic expressions, though pragmatically significant, are semantically nonsignificant.

2.3. *Main Theorems*

Firstly an immediate consequence of Definition 1:

COROLLARY 4.1 For any constructs P, Q, R such that $P=Q \,\&\, R$ is defined,

$$\mathcal{I}(P) \supseteq \mathcal{I}(Q), \mathcal{I}(R).$$

Example By definition, $x<y = x \leqslant y \,\&\, x \neq y$. Since "$\neq$" equals "not $=$", $\mathcal{I}(\neq) = \overline{\mathcal{I}(=)}$ is nonempty. Hence $\mathcal{I}(<) = \mathcal{I}(\leqslant) \cup \mathcal{I}(\neq) \supseteq \mathcal{I}(\leqslant)$. Briefly, "$\leqslant$" is intensionally poorer than "$<$".

THEOREM 4.1 If P and Q are either predicates or statements, and $P \vee Q$ is defined,

$$\mathcal{I}(P \vee Q) = \mathcal{I}(P) \cap \mathcal{I}(Q).$$

Proof By logic $P \vee Q = \neg(\neg P \,\&\, \neg Q)$. By Definition 1(iii), $\mathcal{I}(P \vee Q) = \mathcal{I}(\neg(\neg P \,\&\, \neg Q))$. By (ii), the RHS is $\overline{\mathcal{I}(\neg P \,\&\, \neg Q)}$. By (i) the last formula equals $\overline{\mathcal{I}(\neg P) \cup \mathcal{I}(\neg Q)}$. Finally, by (ii) we get $\overline{\overline{\mathcal{I}(P)} \cup \overline{\mathcal{I}(Q)}}$ which, by one of de Morgan's theorems in the algebra of sets, gives the desired result.

Example Elucidate possibility of the conceptual kind as $\Diamond p =_{df} p \vee q$ with q indeterminate. Then $\mathcal{I}(\Diamond p) = \mathcal{I}(p) \cap \mathcal{I}(q) \subseteq \mathcal{I}(p)$. That is, possibility statements are intensionally included in their "extensional" bases – which is one reason for avoiding them (whenever possible!).

COROLLARY 4.2 If $\mathcal{I}(P) \subseteq \mathcal{I}(Q)$, then $\mathcal{I}(P) = \mathcal{I}(P \vee Q)$.

Proof By the algebra of sets $\mathcal{I}(P) \subseteq \mathcal{I}(Q)$ iff $\mathcal{I}(P) \cap \mathcal{I}(Q) = \mathcal{I}(P)$. Apply Theorem 1 to the LHS of the last equation.

Remark This result shows that the converse of axiom (iii) is false: cointensiveness does not imply identity. For, if it did, we should have $P = P \vee Q$. This suggests that our concept of intension does not coincide with that of sense. We shall see in Ch. 5 that the former is included in the latter.

THEOREM 4.2 If P and Q are either predicates or statements, then

$$\mathscr{I}(P \Rightarrow Q) = \overline{\mathscr{I}(P)} \cap \mathscr{I}(Q) = \mathscr{I}(Q) - \mathscr{I}(P) \subseteq \mathscr{I}(Q).$$

COROLLARY 4.3 If P and Q are either predicates or statements, then

$$\mathscr{I}(P \Leftrightarrow Q) = \mathscr{I}(Q) - \mathscr{I}(P) \cup \mathscr{I}(P) - \mathscr{I}(Q) \equiv \mathscr{I}(P)\ \Delta \mathscr{I}(Q)$$

where 'Δ' designates the symmetric difference or Boolean sum.

We shall use this result in Sec. 2.6 in relation to differences in intension. It will turn out that "$\mathscr{I}(P)\ \Delta \mathscr{I}(Q)$" is a good measure of the intensional difference between P and Q. Hence while the extension function $\mathscr{E}$ obliterates the differences between equivalents, the intension function $\mathscr{I}$ brings those differences into the open: see Figure 4.1

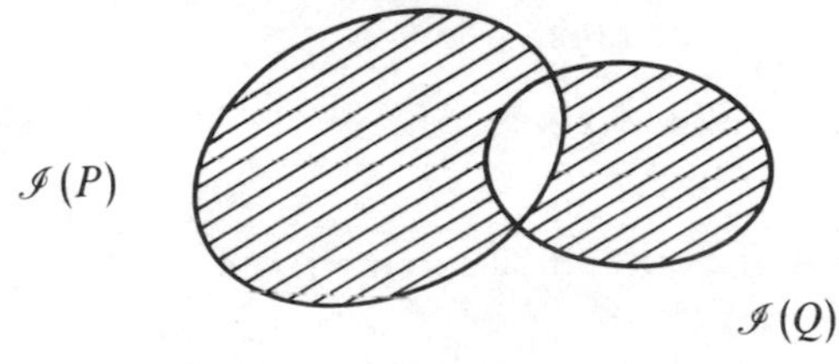

Fig. 4.1. The intension of an equivalence equals the (symmetric) difference between the intensions of the members of the equivalence (Corollary 3).

THEOREM 4.3 For any two predicates (or statements) P and Q, if $P \Rightarrow Q$ is defined then

$$\mathscr{I}(P \,\&\, (P \Rightarrow Q)) = \mathscr{I}(P \,\&\, Q).$$

Proof By Definition 1(i) and Theorem 2,

$$\mathscr{I}(P \,\&\, (P \Rightarrow Q)) = \mathscr{I}(P) \cup (\overline{\mathscr{I}(P)} \cap \mathscr{I}(Q)) =$$
$$= (\mathscr{I}(P) \cup \overline{\mathscr{I}(P)}) \cap (\mathscr{I}(P) \cup \mathscr{I}(Q)) \text{ by distributivity.}$$

The first factor equals U and its intersection with the second factor equals the latter, which by Definition 1(i) in turn equals $\mathscr{I}(P \,\&\, Q)$.

THEOREM 4.4 For any predicate (or statement) P in U

(i) $\mathscr{I}(\neg\neg P) = \mathscr{I}(P)$;

(ii) $\mathscr{I}(P \vee \neg P) = \emptyset$;

(iii) $\mathscr{I}(P \,\&\, \neg P) = U$.

A generalization of Theorem 4(ii) is

THEOREM 4.5 Tautologies are intensionally void.

Proof Consider an arbitrary construct P and a tautologous one T. Because the collection of intensions is an algebra of classes, intersections determine greatest lower bounds or infima. Hence in particular $\inf\{\mathscr{I}(T), \mathscr{I}(P)\} = \mathscr{I}(T) \cap \mathscr{I}(P)$. By Theorem 1 the RHS equals $\mathscr{I}(T \vee P)$. Since the least intension is $\emptyset$, we have $\mathscr{I}(T \vee P) = \emptyset$. And, since the intension of P is comprised between $\emptyset$ and U, we conclude that $\mathscr{I}(T) = \emptyset$.

An obvious consequence of this theorem is that logic cannot alter contents. More precisely, we have

COROLLARY 4.4 Tautologies, if conjoined, do not add; if disjoined they deplete. I.e., for any predicate (or statement) P in U and any tautologous construct T of the same rank,

(i) $\mathscr{I}(P \,\&\, T) = \mathscr{I}(P)$
(ii) $\mathscr{I}(P \vee T) = \emptyset$

Proof By Definition 1 and Theorems 1 and 5.

COROLLARY 4.5 Contradictions, if conjoined, blow up; if disjoined they change nothing:

(i) $\mathscr{I}(P \,\&\, \neg T) = U$
(ii) $\mathscr{I}(P \vee \neg T) = \mathscr{I}(P)$.

Proof Similar to the previous proof.

COROLLARY 4.6 Repetition is futile:

$$\mathscr{I}(P \,\&\, P) = \mathscr{I}(P \vee P) = \mathscr{I}(P).$$

Consequently 'P', '$P \,\&\, P$' and '$P \vee P$' are not different concepts but different signs representing the same concept.

Finally a couple of results concerning the relations in intension between constructs that hold logical relations.

THEOREM 4.6 Whatever entails contains: If P entails Q, then P contains Q in intension and conversely:

$$\vdash P \Rightarrow Q \quad \text{iff} \quad \mathscr{I}(P) \supseteq \mathscr{I}(Q).$$

Proof First the left to right implication. By Theorem 2, $\mathscr{I}(P \Rightarrow Q) = \overline{\mathscr{I}(P)} \cap \mathscr{I}(Q)$. But by hypothesis the conditional is analytic and, by Theorem 5, analytic conditionals are intensionally empty. That is, $\overline{\mathscr{I}(P)} \cap \mathscr{I}(Q) = \emptyset$, which amounts to $\mathscr{I}(P) \supseteq \mathscr{I}(Q)$. Now the converse, i.e.: If $\mathscr{I}(P) \supseteq \mathscr{I}(Q)$ then $\vdash P \Rightarrow Q$. Assume the antecedent. Then, since $\mathscr{I}$ is by hypothesis a one to one map, there is a construct R such that $P = Q \& R$. And by logic $P \vdash Q$.

COROLLARY 4.7 If two constructs are analytically equivalent then they are coextensive and conversely:

$$\vdash P \Leftrightarrow Q \quad \text{iff} \quad \mathscr{I}(P) = \mathscr{I}(Q).$$

Contingent biconditionals, whether formal or factual, do not enjoy the same privilege – hence nonanalytic equivalents are not everywhere substitutable *salva significatione*. For example, in Euclidean geometry the propositional schemata "x is an equilateral triangle" and "x is an equiangular triangle" are equivalent but not analytically so – hence the two predicates are not cointensive. There would be little point in keeping the two concepts, as we do, if they were cointensive. As noted with reference to Corollary 3, the intension of an equivalence consists in whatever is not common to the equivalents. Coextensiveness is a bridge not a gate. This is a reason for not couching definitions, which are identities, in the form ⌜A iff B⌝. We shall return to this question in Ch. 10, Sec. 2.2.

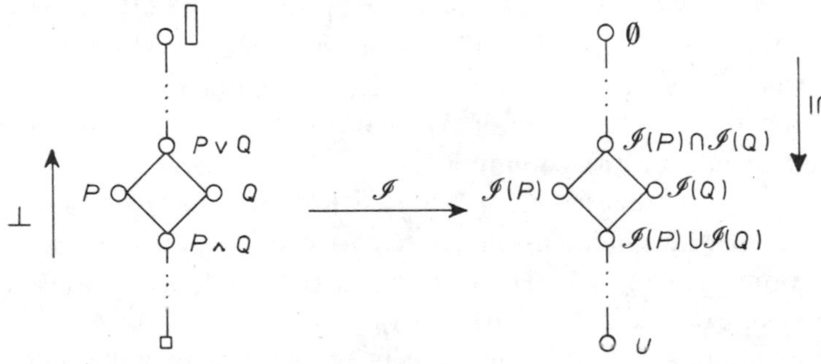

Fig. 4.2. The lattice of predicates and the lattice of intensions: anti-isomorphic.

Definition 1 and the preceding theorems and their corollaries show that there is a *duality*, or anti-isomorphism, between constructs and their intensions, in the sense that $\mathscr{I}$ carries meets into joins and conversely. Look at Figure 4.2.

2.4. *Intensional Difference and Family Resemblance*

The poorer the intension of a construct the more general it is. In other words, the more general (or less specific) of two constructs is the one with the smaller intension. Thus "atom" is more general than "helium atom" precisely because it is the weaker of the two. The difference in intension between a species concept S and a genus concept G, i.e. their *differentia specifica*, consists in whatever characterizes S but not G. Since in our semantics intensions are sets, it is natural to exactify the notion in this way:

$$\delta(S, G)=\mathscr{I}(S)-\mathscr{I}(G)\equiv\mathscr{I}(S)\cap\overline{\mathscr{I}(G)}.$$

The generalization of this concept to an arbitrary pair of constructs calls for the notion of symmetric difference, or Boolean sum, of sets. In this way we assure that the richer or stronger construct will show through. More explicitly, we introduce

DEFINITION 4.8 Let P and Q be either predicates or propositions. Then the *difference in intension* between them is the symmetric difference between their individual intensions:

$$\delta(P, Q)=\mathscr{I}(P)\,\Delta\,\mathscr{I}(Q)\equiv\mathscr{I}(P)-\mathscr{I}(Q)\cup\mathscr{I}(Q)-\mathscr{I}(P).$$

Example Let T and T' be two axiomatized theories with axiom bases (taken conjunctively) A and $A'=A\ \&\ B$ respectively. T' is the stronger of the two theories, i.e. the one with the richer intension, whereas T is the more general (less specific). For example, T could be the general theory of groups and T' the theory of Abelian groups: in this case the extra axiom B would be the commutative law. The difference in intension between the two theories is $\delta(T, T')=\mathscr{I}(A\&B)\,\Delta\,\mathscr{I}(A)=\mathscr{I}(B)\cap\overline{\mathscr{I}(A)}$. This is the newness in intension, relative to A, contributed by the additional axiom. This enrichment in sense is balanced by a shrinking in coverage or extension: $\mathscr{E}(A\&B)=\mathscr{E}(A)\cap\mathscr{E}(B)\subseteq\mathscr{E}(A)$.

Let us apply the newly won concept to the case of independent or "perpendicular" constructs characterized by Definition 6:

THEOREM 4.7 The difference in intension between two intensionally independent construct equals the union of their separate intensions:

$$\text{If } P \perp Q, \text{ i.e. } \mathcal{I}(P) \cap \mathcal{I}(Q) = \emptyset, \text{ then } \delta(P, Q) = \mathcal{I}(P) \cup \mathcal{I}(Q).$$

Proof By Definition 8 and recalling that, if A and B are disjoint sets, then $A \cap \bar{B} = A$.

Example Aristotle's metaphysical categories (substance, quality, time, etc.) are intensionally disjoint. Hence any two of them differ in whatever both connote.

Corollary 3 can now be reworded as

THEOREM 4.8 The intensional difference between two constructs equals the intension of their equivalence:

$$\delta(P, Q) = \mathcal{I}(P \Leftrightarrow Q).$$

Proof By Corollary 3 and Definition 8.

Although equivalences obliterate extensional differences, they bring out differences in intension. This was to be expected in view of the anti-isomorphism between constructs and their intensions noted at the end of Sec. 2.3. Indeed, if the logical difference between two constructs is exhibited by their inequivalence, their semantical difference must be given by their equivalence. In other words, the function δ defined by Definition 8 does the opposite of its counterpart, the function $d: U \times U \to U$ that takes pairs of constructs into their inequivalences, i.e. such that

$$d(P, Q) = (P \,\&\, \neg Q) \vee (P(\neg P \,\&\, Q) = P \not\Leftrightarrow Q \text{ for } P \text{ and } Q \text{ in } U.$$

Furthermore any mapping $f: U^n \to U$ preserving all the values of the *logical* distance d may be regarded as a *tautological transformation*. On the other hand any mapping $g: U^n \to U$ that preserves all the values of the *semantical distance* δ may be interpreted as a transformation (whether logically valid or not) that does nothing to bring the constructs either closer together or farther apart. We shall not make a detailed investigation of the algebraic structure given to U by δ in analogy to the one induced by d on U. (For the latter see Blumenthal and Menger 1970.) For our purposes it will be enough to prove

THEOREM 4.9 Let U be a Boolean algebra of either predicates or statements. Then the family of their intensions, i.e.

$$\mathbb{I}=\{\mathscr{I}(P) \mid P\in U\}=\{\mathscr{I}(P)\in\mathscr{P}(U)\}$$

is a ring of sets to be called the *ring of intensions* of U.

Proof To begin with $\mathbb{I}$ is not empty: even if all the P's in U are tautologies, $\mathbb{I}$ contains the empty set. Next, by Definition 1, the union of any two intensions is an intension – namely the intension of a conjunction. Likewise for the differences. Consequently the intersection of two intensions is an intension. Hence $\mathbb{I}$ is a ring of sets.

This theorem is just a formal prerequisite for the following investigation, which is of interest to semantics. The set valued function δ is a non-numerical measure of the "distance" in intension between two constructs. This is no metaphor: we shall presently show that δ is formally similar to the distance function in a one dimensional quasimetric space. (While in a metric space two points are the same just in case they are not separated, in a *quasimetric* space distinct points need not be separated. In these spaces the weaker condition $\delta(x, x)=0$ holds instead of $\delta(x, y)=0$ iff $x=y$.) We shall call *pseudoquasimetric space* any nonempty set equipped with such a metric δ. We proceed to prove that intensions form such a space.

THEOREM 4.10 Let $\mathbb{I}$ be a ring of intensions. Then the structure $\langle\mathbb{I}, \delta\rangle$, where $\delta:\mathbb{I}^2\to\mathbb{I}$ is the difference in intension, is a pseudoquasimetric space, i.e. the function δ satisfies all of the following requisites:

(i) $\delta(P, Q)\supseteq\emptyset$;
(ii) $\delta(P, Q)=\delta(Q, P)$;
(iii) $\delta(P, Q)\,\Delta\,\delta(Q, R)=\delta(P, R)$;
(iv) $\delta(P, P)=\emptyset$

for any P, Q, R in the substrate U of $\mathbb{I}$.

Proof All four properties but the third are easily checked by recalling Definition 8. The "triangle equality" (iii) follows from the associativity of the Boolean sum and the equations: $A\Delta A=\emptyset$ and $A\Delta\emptyset=A$.

The interest of this theorem is twofold. First, it reinforces our decision to regard "$\mathscr{I}(P)\,\Delta\mathscr{I}(Q)$" as the difference or distance in intension between P and Q. We may now picture intensions as points on a line: see

Figure 4.3. Second, we can now define neighborhoods,

$$\mathscr{I}(P) \leftarrow \delta(P, Q) \rightarrow \mathscr{I}(Q)$$

Fig. 4.3. The intension space is one-dimensional

that will prove to be both mathematically and philosophically significant:

DEFINITION 4.9 Let $\mathbb{I}$ be a ring of intensions over an algebra U of constructs and call A a fixed subset of $\mathbb{I}$. Then for P in U we call

$$N_A(P) = \{\mathscr{I}(Q) \mid Q \in U \quad \text{and} \quad \delta(P, Q) \subset A\}$$

the *A-neighborhood* of P.

This concept of neighborhood in the space of intensions elucidates the fuzzy notion of *family resemblance* which the second Wittgenstein and his followers made so much of. Indeed, the A-neighborhood of a construct P is the collection of intensions such that their distance to the intension of P falls within the given intension interval A. Every $N_A(P)$ is thus a collection of constructs near P. The smaller A the closer the kinship.

Furthermore the neighborhoods introduced by Definition 9 generate two different topologies. More precisely, we can prove

THEOREM 4.11 The collections of neighborhoods

$$\mathscr{B}_1 = \{N_A(P) \mid P \in U \quad \text{and} \quad A \subset \mathbb{I}\}$$
$$\mathscr{B}_2 = \{N_A(P) \mid P \in U \quad \text{and} \quad A \in U\}$$

each constitute a base for a topology on U. That is, in each case (*a*) the union of neighborhoods equals U; (*b*) if a construct P belongs to the intersection of two given neighborhoods, then there is a neighborhood of P that is included in the intersection.

By virtue of the first clause of this theorem, no construct in a given universe of discourse is isolated, and the various families of constructs cover the entire universe. By virtue of the second clause any construct belonging to two families can be placed in a third family included in the overlap of the former and therefore constituting a set of close relatives. We shall not pursue this line here although it sounds promising because topology – rather than natural language – is the natural tool for refining concepts of vicinity and togetherness.

In closing we note that the semantic concept of family resemblance is not reducible to pragmatic (e.g. psychological or linguistic) concepts. Thus the semantic distance between two constructs is determined by the theory in which they occur not by the way they are conceived or mentioned by a subject under certain circumstances. In particular, semantic distance is unrelated to both psychological association and linguistic correlation. Whether or not two constructs are semantically close in a given context, they may be strongly correlated for some subjects and uncorrelated for others, particularly if they have never thought of them before. This being so, the attempts to account for semantic sense in psychological or in linguistic terms are bound to fail.

3. Some relatives – kindred and in law

3.1. *Logical Strength*

Our concept of intension is coextensive with that of logical strength. In fact, by Theorem 6 the stronger a construct the richer it is and conversely: $P \vdash Q$ iff $\mathscr{I}(P) \supseteq \mathscr{I}(Q)$. Moreover our calculus of intensions satisfies a number of conditions characterizing the notion of logical strength or deductive power. In particular, the desiderata D4, D5, D7, D8 and D9 listed in Sec. 2.1 are fulfilled by the two notions. In turn, the concept of logical strength coincides with that of specificity: the more specific (less generic) a construct the stronger it is. Thus the concept of reading skill is logically stronger and intensionally richer than the generic concept of skill. Consequently a theory of reading skill would be more specific (stronger) than a general theory of skills.

Although the concepts of strength and of intension are coextensive, they are not cointensive. (Logical strength is cointensive with import, to be studied in Ch. 5, Sec. 3.) In fact our Definition 1, which determines the intension of "intension", does not contain the concept of deducibility, which is on the other hand the very kernel of the concept of logical strength, and which defines both Carnap's and Popper's concepts of strength. This situation exemplifies our thesis that coextensives need not be cointensive or, for that matter, not even coreferential. A simpler example is this: ⌜Ormazd is unhappy iff Ahriman is happy⌝. A more sophisticated example: "consistency" and "satisfiability" (or "validity in a model") are coextensive, whence finding a model of a set of formulas proves the latter's consistency; nevertheless the syntactic notion is not cointensive with the semantic one.

We now turn to a few other notions that are often mixed up with that of intension.

3.2. *Information*

Intension is often equated with information content. Thus it is frequently asserted that, whereas synthetic propositions convey information, analytic formulas do not. No doubt there is a relation between intension and information: the greater the former the richer the latter. However, the two are not identical. To begin with a construct *has* a content whereas a signal, such as a written sentence, *conveys* (carries, transmits) information *to someone*. In other words, whereas intension is defined on the set of constructs, information is defined on the set of pairs signal-subject, where a subject is a viewer or hearer competent to decode the signal. Information, to be received, requires a receiver, i.e. a system equipped with suitable receiving and decoding devices. The sentence '$(-1)^2=1$', which expresses or conveys the statement that the square of minus one equals unity, transmits no information to a baby and it may convey the wrong information to a mathematician under the influence of LSD. In short, whereas the concept of intension is semantical, that of information is pragmatical: the former is subject free, the latter is tied to a subject.

Upon shoving aside the all important information receiver and the no less important information channel, it is possible to build an impersonal or semantic concept of information. In fact, upon removing the concept of a subject from the foregoing considerations and rendering them more explicit, we are left with the following principles.

INF 1 If a signal (mark, sign, inscription, sound, etc.) is a sentence or represents a sentence, then the *information conveyed by the signal* is the proposition designated by the sentence.

INF 2 If S and S' are sets of signals representing the sets of propositions P and P' respectively, then

(i) the information conveyed by S is *larger than or equal to* the one conveyed by S' iff $P \supseteq P'$;

(ii) the *information gain* accompanying the substitution of S for S' equals $P - P' \equiv P \cap \overline{P'}$.

These three propositions define what will be called our *semantic concept of information*. It boils down to this: The information or message conveyed by a signal consists of the proposition or propositions the signal stands for. It follows (*a*) that nonpropositional signals convey no information, (*b*) that the greater the content of a proposition the richer the information carried by the signal representing that proposition, and (*c*) that the truer a proposition the more accurate the information carried by the signal representing that proposition. These platitudes may be baptized *Simple Simon's semantic view on information.*

Simple Simon's view contrasts with the quantitative theories of semantic information proposed over the past two decades, from that of Bar-Hillel and Carnap (1953) to Hintikka's (1968, 1970). In particular, Simple Simon's view is at variance with the former, which initiated what is today a bulky body of literature on the subject. This theory boils down to identifying the amount of information conveyed by a statement s with its improbability:

$$cont\ s = 1 - Pr(s).$$

(For a similar proposal see Popper, 1963b.) Once this formula is accepted the whole of the elementary theory of probability can be rewritten in terms of the *cont* function and exploited for semantic purposes. The outcome is as tidy as the probability calculus, of which it is a rewrite – but it is, alas, hollow. Indeed the *cont* function can hardly be said to exactify any intuitive concept of content. For one thing contradictions turn out to have maximal content, while alternations such that the probabilities of the disjuncts are complementary are assigned zero content. For another the theory assigns coextensives the same probability, hence the same content, and is therefore at a loss to distinguish ⌜The grass is green⌝ from ⌜The weather has been mild and wet⌝, which do not even have the same referent. Worse, the theory is not applicable, because neither it nor inductive logic (with which it is sometimes associated) tell us how to assign probabilities to the constituent statements – nor even how to interpret the expression '$Pr(s)=p$'. Whence there is no point in claiming that every formula, for example the Schrödinger equation, conveys a definite amount of information lying between 0 and 1. Thou shalt not take the name of the Number thy God in vain.

Other quantitative theories of information are paraphrases of, or heavily indebted to, the much talked about but little understood mathematical theory of communication, or statistical information theory (Shannon and Weaver, 1949). The problem this theory set out to solve was to identify and measure the information carried by physical signals distorted by random perturbations causing noise. The basic notion in this theory is that of the probability of a random binary physical signal – not of a proposition or statement. This probability is a number that can be estimated provided something is known or assumed about the source-channel-receiver-environment system. It is, let us emphasize, the probability of occurrence of a *random physical event*, not the probability of a construct, let alone an orderly one such as a scientific hypothesis. The theory postulates that the amount of information carried by such a signal s_i, which pops up with probability p_i at the receiving end of a communication channel, depends only on this probability, namely thus:

$$I(s_i) = -\log_2 p_i \text{ bits.}$$

Consequently the average amount of information H conveyed by a string of N mutually independent random signals is given by Shannon's formula

$$H = \sum_{i=1}^{N} p_i I(s_i) = -\sum_{i=1}^{N} p_i \log_2 p_i \text{ bits/signal, with } \sum_{i=i}^{N} p_i = 1.$$

If all N signals are equally probable then $p_i = 1/N$, whence $H = \log_2 N$ bits/signal. In information theory H is often interpreted as a measure of the *unexpectedness* or *news value* of the message. Or, what is equivalent, the amount of information measures the ignorance or uncertainty of the receiver concerning the signal about to be received. All these are pragmatic or psychological concepts not semantic ones. Not even Kolmogoroff's generalization of Shannon's formula is relevant to semantics, as his information theoretic "entropy" is defined only for random variables, i.e. for variables associated with some probability distribution.

The preceding information theoretic concepts are far removed from the semantic concepts of intension and content. For one thing the concept of randomness, central to any application of probability theory, makes no sense with reference to propositions. (The idea of stumbling on a given proposition with a certain probability does make sense but only in the context of pragmatics – and in any case bears no relation to the content of the proposition.) For another the information theoretic concept of amount of information is in a sense the very opposite of the semantic concept of content, inasmuch as a signal conveying a proposition with a definite and rich content, such as a law of conservation of energy, has zero information content for a physicist, as it comes as no surprise to him. For all these reasons the statistical theory of information is irrelevant to semantics.

In view of the fundamental difficulties met by the quantitative theories of semantic information utilizing the statistical information theory, we shall stick by Simple Simon's concept of information. However we find no use for it precisely because it is a semantic concept, whereas information and communication belong in pragmatics. Whether or not a sign carries information to someone depends on the receiver's background knowledge and thirst for new knowledge as much as on the content of the statement being transmitted. Thus a tautology new to a given subject is informative to him, whereas a factual sentence he knows by heart is not. In any case neither the semantic concept of information nor any of the pragmatic concepts of information can be equated with the concept of intension.

3.3. *Testability*

Peirce, Frege, the Vienna Circle and the first Wittgenstein held the identity of meaning and verifiability. Carnap transformed this thesis into the equation of meaning with test procedures: "the meaning of a sentence is in a certain sense identical with the way we determine its truth and falsehood; and a sentence has meaning only if such a determination is possible" (Carnap, 1936). This thesis was subjected to such devastating criticisms (see, e.g., Williams, 1937; Russell, 1948; and Hempel, 1965) that it has now been almost abandoned by philosophers. (Quine, 1971, is about the only faithful left.) However, it survives among scientists and even, in a watered-down form, in Popper's thesis that "the degree of testability of a statement increases with its content" (Popper, 1963a, 1963b). *Prima facie* this thesis is plausible enough: tautologies have no content and are insensitive to empirical tests, whereas factual statements have a content and are presumably susceptible to such tests. Moreover, a disjunction is both intensionally poorer and less vulnerable to experience than either of its disjuncts, and "It is possible that *p*" says less and therefore risks less than "*p*".

It is difficult to assess the above thesis in the absence of a full fledged theory of degrees of testability (not just *a posteriori* confirmation) in relation to content – not to speak of a theory of content. However, the thesis is ruined by the following counterexamples. A singular observational statement such as "That apple is red" is both intensionally indigent and eminently testable. On the other hand any of the partial differential equations of physics has such a rich content that it is not fully testable. In fact one can check only a few of its (infinitely many) solutions at some of the argument values. Of course such a rich statement has more opportunities of being tested and clashing with data than a scientifically trivial statement such as "That apple is red", which can be checked at a glance. That is, the richer the intension of a statement the more tests it *calls for* in order to find out its truth value, i.e. the greater the target it offers to experience. But this can hardly be called 'testability'. It might be more aptly be called *un*testability: the more a statement "says" (the richer its sense) the more there is to be checked, hence the less fully can this task be performed. This capacity of a statement to come to grips, or making contact, with experience (Popper's 'testability') we shall call *empirical sensitivity* (or target or cross section). The empirical sensitivity of a statement (or of a collection of statements) is, intuitively speaking, inversely related to its testability proper, or capacity of being checked for truth.

This is not the place to elaborate on this problem. Therefore let me express telegraphically my views on the relations between sense and the methodological concepts we are discussing.

(i) The concepts of sense, empirical sensitivity, and testability, are *heterogeneous*, hence not mutually reducible. The first is semantic whereas the other two are methodological. The determination of sense is a theoretical affair and moreover it precedes the determination of both empirical sensitivity and testability. Besides, the degree of testability of a statement must be assessed in relation to a host of items, both theoretical and empirical, in the background knowledge.

(ii) The *empirical sensitivity* (target, cross section) a statement offers to data increases with its sense.

(iii) The *testability* of a statement is inversely related to its sense and increases with its systemicity, i.e. with the strength of the ties it holds to other members of a body of knowledge. (A strictly stray statement would be untestable. A statement belonging to a theory that partially overlaps with another theory may have the chance of being double checked: directly and via the other theory.)

(iv) The *degree of confirmation* (empirical validation, corroboration) of a testable statement decreases with its sense. Unlike empirical sensitivity and testability, which are estimated prior to tests, degrees of confirmation are assigned, if at all, *a posteriori*. And they bear no fixed relation to either empirical sensitivity or testability; in particular, a highly testable statement may prove utterly false. On the other hand degrees of confirmation depend critically upon test procedures: different techniques will corroborate (or infirm) a hypothesis to different degrees.

4. Concluding Remarks

Our theory of intensions boils down to the calculus in Sec. 2. This calculus enables one to clarify a number of obscure notions, such as those of intensional inclusion and intensional independence. We shall see in Ch. 9, Sec. 1.6, that our theory also allows us to state and even prove the reciprocal relation between intensions and extensions. Moreover, it enables us to compute the sense of a whole as a function of the senses of its parts – provided we either know the latter or do not care too much for them, for remaining satisfied with finding intensional relations. This shortcoming will be remedied in part in the next chapter, where we shall learn to find the full sense of a theoretical construct. It will turn out that the intension of a construct is included in its full sense.

Our theory construes intensions as basic or irreducible semantic objects. It is couched in set theoretic terms but it does not reduce intensions to either extensions or reference classes. Nor does our theory make use of modal concepts. In particular, our Definition 4 is at variance with the following construals of the notion of intensional inclusion (Lambert and van Fraassen, 1970):

> P is intensionally included in Q iff ⌜Necessarily, all the individuals which are P are Q⌝ is true.

and

> P is intensionally included in Q iff ⌜All (possible) individuals which are P are Q⌝.

We have kept our semantics free from modalities and independent of modal logics for several reasons. Firstly because we do not need them. Secondly because it is far from clear how modal prefixes should be interpreted. If all 'necessarily' is intended to mean is logical necessity, then the concepts of entailment do this job and much better: they elucidate a notion of relative (not absolute) necessity – the necessity of a conclusion relative to its premises and to the accepted rules of inference. And if what is meant is ontic (or physical) necessity, then it has no place in semantics – nor, for that matter, does modal logic supply an adequate elucidation of that concept. In sum, I find no use for modal logic in semantics. On the other hand semantics should be used to try and solve some of the riddles of modal logic, such as whether or not the statement ⌜That trick may work⌝ means the same as ⌜That trick may not work⌝.

So far we have looked into one of the dimensions of sense. We now turn to the remaining dimensions – purport and import.

CHAPTER 5

GIST AND CONTENT

Our analysis of sense in the previous chapter has been "local" or "horizontal": it was limited to the given construct with little regard for its logical relatives. The limitations of such an approach are obvious. For one thing it cannot do justice to propositions of the form ⌜A means B⌝, where 'means' stands for "entails" or for "is entailed by". For example, that x is loved "means" (follows from) somebody loving x; it also "means" (entails) that x is lovable. In the present chapter we shall supplement the "horizontal" or "local" approach to sense with a "vertical" or "global" analysis. We shall in fact ask ourselves the questions: What is the ancestry (or set of implicants) of a construct?, and What is the progeny (or set of implicates) of a construct? In other words, we will elucidate the notions we have called *purport* and *import*.

We shall formalize the ideas (*a*) that the purport of a construct in a given context is the collection of constructs upon which it depends, or which determine it (logically), and (*b*) that the import of a construct in a given context is the collection of constructs that hang from it, or that are determined by it (logically). The concepts of purport and import are thus mutually dual. And both are context-dependent: the purport and the import of a construct depend on the body of knowledge in which it occurs. This relativization of sense to context curtails a certain freedom – the freedom to make arbitrary meaning shifts. The advantage of such a relativization is clear: it is well-nigh impossible to determine the exact sense of a stray predicate – whence the endless semantic disputes in the young growing fields as well as in the old undisciplined areas. Only systemic predicates and statements, i.e. constructs belonging to definite deductive systems, have definite purports and imports. (When transplanted to a different theory, if not rejected, a construct may well acquire a new sense – i.e. it may become a somewhat new construct.)

Finally the full sense of a construct in a given context may be taken to equal the union of its purport and its import in that context. Consequently the intension or "inner sense" of the construct will be included in its full sense. But, since enjoying a definite place in an axiomatic pecking order is the exception rather than the rule, we shall be well advised to keep the theory of intensions even if intensions are only parts of full senses.

1. Closed contexts

1.1. *Closed Contexts and Their Structure*

Recall the notion of a context introduced in Ch. 2, Sec. 3.4. A context $\mathbb{C}=\langle S, \mathbb{P}, D\rangle$ is a construct composed of a set S of statements, the extralogical predicates occurring in which belong to a family $\mathbb{P}$ of predicates all of which refer to individuals in a domain D of objects.

A context may be amorphous or structured. See Table 5.1.

TABLE 5.1

Conceptual systems

Item	Structure	Semantics
Set of constructs	———	———
Context	———	Referential homogeneity
Closed context	Boolean algebra	Referential homogeneity
Theory	Filter	Referential homogeneity
Consistent theory	Ultrafilter	Referential homogeneity

We get a structured context if we keep D and $\mathbb{P}$ fixed and allow all and only logical operations in the sets $\mathbb{P}$ and S. Such a context will be closed both formally (syntactically) and as regards reference (semantically). The former because logical processing will produce nothing outside S, the latter because no referents foreign to D will be allowed to intervene in the course of that processing. (Such a double closure need not stifle research: we are always at liberty to leap to a different context.) More explicitly, we lay down

DEFINITION 5.1 The structure $\mathbb{C}=\langle S, \mathbb{P}, D\rangle$ is called a *closed context* iff (i) $\mathbb{C}$ is a context and (ii) S is closed under negation, conjunction, disjunction, and generalization (both existential and universal).

From an algebraic point of view the statements in a closed context constitute a complemented lattice. Indeed every member s of S has its opposite number $\neg s$ in S and, for any two elements s and t of S, both the join $s \vee t$ and the meet $s \wedge t$ (i.e. the conjunction $s \& t$) are in S. Moreover $\vee$ distributes over $\wedge$ and conversely, so that the lattice is distributive in addition to being complemented. And because of the latter property S contains a null element □ as well as a universal element ▯. Every extensionally void statement equals □ and every extensionally universal statement equals ▯. In brief, we have

THEOREM 5.1 The statements in a closed context constitute a distributive complemented lattice with zero element and unit element – i.e. a Boolean algebra.

A similar result holds for certain subsets of the collection of predicates in a closed context. To prove this we recall how the basic connectives are defined for predicates (Ch. 1, Sec. 2.2).

DEFINITION 5.2 Let $\mathbb{C} = \langle S, \mathbb{P}, D \rangle$ be a closed context. If P and Q belong to $\mathbb{P}$, and are both defined on a subset $E \subseteq D$, then

(i) $\neg P : E \rightarrow S$, with $\neg Px = \neg(Px)$ for every x in E;

(ii) $P \wedge Q : E \rightarrow S$, with $(P \wedge Q)\, x = Px \wedge Qx$ for every x in E.

Consider now the totality of predicates in $\mathbb{P}$ with domain $E \subseteq D$, i.e. $\mathbb{P}_E = S^E$. Since by hypothesis S is a lattice and moreover a Boolean algebra, mathematics tells us that $\mathbb{P}_E = S^E$ has the same algebraic structure. In other words, we have

THEOREM 5.2 The totality of predicates defined on the same domain and belonging to a closed context form a complemented distributive lattice with zero element and unit element – i.e. a Boolean algebra.

By virtue of the algebraic similarity between the set S of statements and the subset $\mathbb{P}$ of predicates in a closed context, we may call the two by the generic name *closed set of constructs*. In this way Theorems 1 and 2 can be lumped into the

PROPOSITION Every closed set of constructs is a Boolean algebra.

This formal unification of statements and predicates will save us ink provided we remember that it is not the whole set $\mathbb{P}$ of predicates in a closed context, but only those subsets of $\mathbb{P}$ constituted by predicates with equal domains, that have the said Boolean structure. For example, "wave length" and "refractive index" belong to wave optics, which is a theory and therefore a closed context, but since they are not defined on a common domain they do not belong to a closed predicate context.

1.2. *The Logical Ancestry of a Construct*

Since the purport of a construct in a context will prove to be the principal ideal generated by the construct, it will pay to recall the ABC of ideals. An ideal in a lattice L is that subset I of L that contains all the predecessors of any given element of L as well as all the joins of any two elements of I (including of course the join of a member with its complement). More explicitly we have the following definition:

If L is a lattice, then I is an *ideal* in L iff I is a nonempty subset of L satisfying the following conditions:

*I*1 For all $x \in I$ and $y \in L$, if $y \leqslant x$ then $y \in I$.
*I*2 For all $x, y \in I$, $x \vee y \in I$.

Clearly every lattice is an ideal. (Even the singleton composed of the null construct is an ideal.) In particular, our closed sets of constructs are ideals. And removing the universal element $\square$ from I turns I into a *proper ideal* $I \subset L$. Finally, making sure that $\square$ is in a proper ideal authorizes us to pronounce it a *maximal ideal*. In other words, a proper ideal is a maximal ideal iff, for every $x \in L$, both x and its complement $\bar{x}$ are in I.

Let us apply these algebraic concepts to our lattices of constructs in closed contexts. Such an application requires only one more step, namely to identify the order relation. If the basic set happens to be one of either predicates or statements, then the order relation is that of entailment $\vdash$. That is, if x and y are in a closed set $\mathbb{C}$ of constructs, then

$$x \leqslant y \quad \text{iff} \quad x \vdash y.$$

It is now easy to check that every closed set of constructs from which the universal construct is missing, is a maximal ideal. For reference:

THEOREM 5.3 Let $\mathbb{C}$ be a closed set of constructs partially ordered by the relation $\vdash$ of entailment. Then $\langle \mathbb{C}, \wedge, \vee, \neg, \square \rangle$ is a maximal ideal.

Every ideal can be analyzed into as many partial ideals (or subideals) as elements in the base lattice. Indeed, every element in the lattice drags the whole lot of its predecessors to form its own private ideal or family tree. More precisely, we have the following definition:

For every element x in a lattice L, the set $\{y \in L \mid y \leqslant x\}$ is called the *principal ideal* generated by x in L. Symbol: $(x)_L$.

The adaptation of these algebraic concepts to our needs yields the following

DEFINITION 5.3 Let x be in a closed set of constructs $\mathbb{C}$. Then the *logical ancestry* of x in $\mathbb{C}$ is the principal ideal generated by x in $\mathbb{C}$:

$$(x)_{\mathbb{C}} = \{y \in \mathbb{C} \mid y \vdash x\}.$$

Note that the ancestry of a construct depends on the context in which it occurs.

We proceed to put the preceding ideas to work in semantics.

2. Sense as Purport or Logical Ancestry

2.1. *Purport and Gist*

The purport of a construct, we said in Sec. 1, may be regarded as the totality of its conceptual determiners, i.e. the collection of constructs that "define" or prove it. A natural explication of this intuitive idea is that of logical ancestry introduced by Definition 3. More explicitly we stipulate

DEFINITION Let x be a construct belonging to a closed set of constructs $\mathbb{C}$. Then the *purport* of x in $\mathbb{C}$ equals the logical ancestry of x in $\mathbb{C}$, i.e. the principal ideal generated by x in $\mathbb{C}$:

$$\mathscr{P}\!ur_{\mathbb{C}}(x) = (x)_{\mathbb{C}} = \{y \in \mathbb{C} \mid y \vdash x\}.$$

For example, because different proofs of a given theorem may exhibit different connections, the purport of a theorem will depend on the premises employed to prove it. This change in purport is corresponsible for the diversity in pragmatic meanings, i.e. for the fact that the same theorem may mean different things to different people (cf. Wang, 1966).

Of all the determiners of any given construct x some may be more basic than others and, by the same token, they may point to whatever is essential in x. In other words, the set of ultimate determiners sufficient to determine x constitute the basic purport or gist of x. More explicitly, we make

DEFINITION 5.5 If x is a member of a closed set of constructs $\mathbb{C}$, then the *gist* (or *basic purport*) of x in $\mathbb{C}$ is the smallest subset of elements of $\mathbb{C}$ that entail x.

For example, the gist of "$\emptyset$" in the algebra of sets consists of two statements: one to the effect that the union of $\emptyset$ with an arbitrary set equals the latter, and the other to the effect that the intersection of $\emptyset$ with an arbitrary set equals the former. We shall return to the concept of gist in the next subsection.

If an ideal containing the null element $\square$ is deprived of the unit element $\Box$, it turns into a maximal ideal. (Recall Sec. 1.2.) Now, removing the universal construct $\Box$ from a closed set of constructs amounts to keeping only the nontautologous constructs, i.e. the extralogical ones. This justifies

DEFINITION 5.6 Let x belong to a closed set of constructs $\mathbb{C}$. Then the *extralogical purport* of x in $\mathbb{C}$ is the maximal principal ideal generated by x in $\mathbb{C}$, i.e.

$$\mathscr{P}ur_{\mathbb{C}\bar{L}}(x)=\{y\in\mathbb{C}\mid y\vdash x \quad \text{and} \quad y\neq\Box\}$$

Remark 1 The extralogical purport of a logical item in $\mathbb{C}=L$ is nil. *Remark 2* If $\mathbb{C}$ is a nonlogical theory (in the sense that it includes not only logical predicates but also extralogical ones), then the extralogical purport of any construct in $\mathbb{C}$ consists of a set of mathematical or factual items. *Remark 3* It is not that tautologies are taken out of the way of deduction in order to find the extralogical purport of some construct: if this were done then deduction would hardly be possible. Tautologies are removed once the deductive process is completed. They are counted as midwives or catalyzers rather than as ancestors of the constructs we are interested in.

Example Consider the most famous, though not the best understood, of all law statements: Newton's law of gravitation $N=\ulcorner F=\gamma mm'/r^2\urcorner$. In the context of Newton's own theory of gravity the law is an axiom and thus constitutes its own purport. But in a wider context, such as the classical theory of gravitation, N is derived from a far more general law (Poisson's) as a special case, namely when the source mass m is concentrated at a point. In this case the mass density can be set equal to $m\delta(r)/r^2$, where δ is the Dirac delta. The skeleton of the deductive tree (i.e. if all the mathematical auxiliaries are discarded) is this:

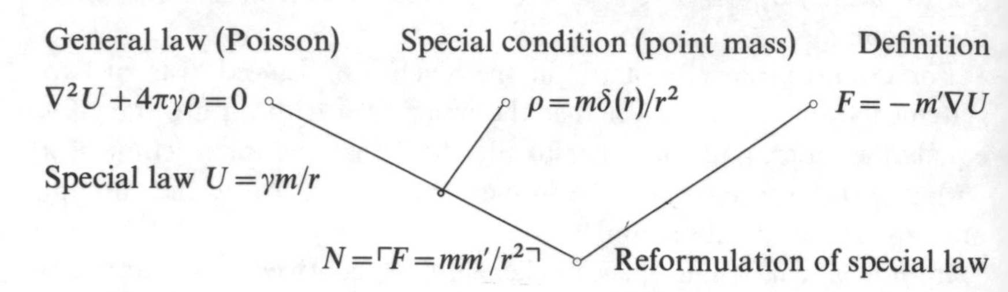

Purport of $N = \{\nabla^2 U + 4\pi\gamma\rho = 0,\ \rho = m\delta(r)/r^2,\ F =_{\mathrm{df}} -m'\nabla U,\ U = \gamma m/r, N\}$ (Actually this, when "read in factual terms", i.e. upon being enriched with the requisite semantic assumptions, is the *factual* purport of Newton's law of gravity: see Sec. 2.3.)

The next desirable step in the development of our theory would be to set up a calculus of purports enabling one to compute, say, $\mathscr{P}ur_{\mathbb{C}}(x \wedge y)$ out of $\mathscr{P}ur_{\mathbb{C}}(x)$ and $\mathscr{P}ur_{\mathbb{C}}(y)$. Yet, it is not apparent how this could be done: there are no obviously regular or lawful relations among the principal ideals in a given lattice. In particular, it is not the case that, for arbitrary elements x, y in a general lattice L, $(x \vee y)$ will equal either $(x) \cup (y)$ or $(x) \cap (y)$. A simple counterexample is this:

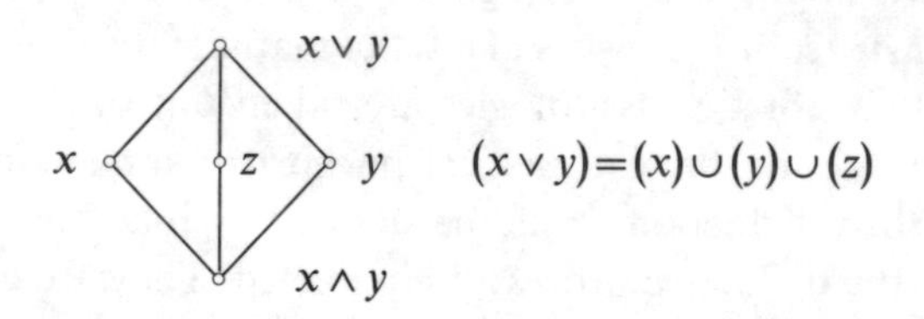

The only apparent simple relations between purports are shown in

THEOREM 4 For any two constructs x, y in a set of closed constructs $\mathbb{C}$,
(i) $\mathscr{P}ur_{\mathbb{C}}(x) = \mathscr{P}ur_{\mathbb{C}}(y)$ iff $x = y$
(ii) If $x \vdash y$ then $\mathscr{P}ur_{\mathbb{C}}(x) \subseteq \mathscr{P}ur_{\mathbb{C}}(y)$.

2.2. *The Gist of a Basic Construct*

Let us now apply the foregoing considerations to the most important constructs in a well organized theory, namely its basic (primitive) concepts and its basic assumptions (axioms). Since such constructs have no logical ancestry other than themselves, they are their own purports and moreover their own gists. More precisely, we have the following

COROLLARY 5.1 In an axiomatized theory the gist of an axiom is the singleton constituted by the axiom itself, and the gist of a primitive concept is the set of axioms in which it occurs.

Example 1 In classical particle mechanics (as axiomatized in Bunge, 1967b) the gist of the primitive concept M of mass is determined jointly by the following postulates:

Mathematical axiom M is an additive function from the set P of particles to the non-negative real numbers.

Factual axiom Newton's second law of motion.

Semantic axiom $M(p)$ represents the inertia of particle $p \in P$.

Example 2 The gist of the concept $\triangleright$ of biological ancestry, a primitive in the theory of evolution as axiomatized by Williams (1970), could be given by the following postulates:

Mathematical axiom $\triangleright$ is a strict partial ordering of the set B [of organisms].

Factual axiom For every x in B: $\neg(x \triangleright x)$.

Semantic axiom $\triangleright \hat{=}$ ancestry relation. [I.e. if x and y are distinct individuals in the set B of organisms, then '$x \triangleright y$' designates $\ulcorner x$ is an ancestor of $y \urcorner$.]

Finally an important qualification. Since independent axioms (i.e. not interdeducible postulates) are separate objects, it makes sense to speak of their separate gists – i.e. of themselves. In some cases it is also meaningful to speak of the gists of distinct basic *concepts*, or primitives. For example, in a meet semilattice $\langle S, \leqslant, \wedge \rangle$ the gist of $\leqslant$ does not overlap with that of $\wedge$, as these two concepts are not only mutually independent but also subject to separate axioms. However, this situation is exceptional: usually an axiom contains two or more primitive concepts that determine one another, so that it becomes impossible to reckon their individual gists even though the concepts themselves are distinct and not interdefinable. In other words, in general mutually independent concepts fail to have disjoint gists. An example will make this clear.

Suppose we enrich a meet semilattice with the join operation $\vee$ to obtain a lattice $\langle S, \leqslant, \wedge, \vee \rangle$. Now the two operations, meet and join, intertwine in such a way that it is no longer possible to disentangle their separate gists. Better: they have no separate gists. In effect, on top of the separate axioms for $\wedge$ and $\vee$, such as those of associativity, we now have the typical absorption laws

$$x \wedge (x \vee y) = x, \quad x \vee (x \wedge y) = x.$$

We can still *distinguish* the two concepts: meets precede joins. But we cannot *disentangle* their individual senses, for they overlap. In other words, the meet and the join operations have no separate gists. Only the *theory* of $\wedge$ and $\vee$ as a whole, to wit, lattice theory, has a definite gist – namely the set of its postulates.

Because the gist of a construct is included in its full purport, which is in turn part of the full sense of the construct, we conclude that *the unit of sense is the whole body of theories in which the construct occurs.* A single statement never suffices and a single theory is often insufficient because a concept may do different things at different places. Unless this point is kept in mind, our thesis that purport is included in sense may be criticized by exhibiting alleged consequences that are counterintuitive (Castonguay, 1972). Thus take a biological theory B containing the assumption that all organisms contain carbon. If one forgets where "carbon" comes from, namely chemistry, one will set $\mathscr{Pur}$ ("carbon") = {"organism"}, which is obviously inadequate. The force of this argument against purport consists in its very weakness – i.e. that it ignores that the concept of carbon is not fully characterized by the biological theory B, which borrows it from chemistry. It is the latter that determines the gist (though not the full sense) of "carbon" as consisting of a set of statements containing *inter alia* the definite description "The element with atomic number 6". Our view is that the full sense of "carbon" is determined by the whole lot of theories containing this concept. B is one of them and it contributes just one of the genuine ingredients of the sense of "carbon", namely the statement that carbon enters the composition of all organisms. In conclusion, a meaning analysis of an isolated construct is bound to yield only a partial, hence misleading result.

Let us now take a closer look at the gist of theories.

2.3. *The Gist of a Theory*

We shall now apply the Definition 5 of gist to a theory, or hypothetico-deductive system. This is a very special kind of closed set of constructs: it is a set of statements glued by the relation of deducibility. (In algebraic terms to be clarified in the next section: A theory is a certain subset of a Boolean algebra of statements, namely that which does not contain the null element but contains all the followers of any given element as well as the meet of any two elements. A theory is, in short, a proper filter on a Boolean algebra of statements. And a consistent theory is an ultrafilter, i.e. a proper filter such that, for any element in the Boolean algebra, either the element or its complement is in the filter but not both.)

To determine the gist of a theory it is best to axiomatize the latter. Indeed, the process of tracing the logical ancestry of any formula in the theory leads to and terminates in some or all of the initial statements of the theory. Among such initial formulas we count not just the specific axioms of the theory but also all of its basic presuppositions, i.e. all the axioms belonging to different contexts and employed in the theory (e.g. as lemmas) without being questioned. The totality of such basic presuppositions may be called the *background* of the theory. For example, ordinary logic and variable portions of mathematics belong to the background of every scientific theory. Many scientific theories presuppose, in addition to such formal theories, other (more fundamental) scientific theories. For example modern solid state theories presuppose quantum mechanics. In sum the search for the gist of a theory must not overlook the background of the theory.

The preceding considerations justify

DEFINITION 5.7 Let T be a theory based on a set A of axioms and on a background B, and call L the set of tautologies included in A or in B. Then

(i) the *gist of* T equals $A \cup B$;
(ii) the *extralogical gist* of T equals $(A \cup B) - L$;
(iii) the *specific extralogical gist* of T equals A.

A theory will have a *factual purport* just in case it includes semantic formulas (in the sense of Ch. 3) specifying what factual items the basic concepts of the theory refer to and which if any traits of the referents those concepts are deemed to represent (whether accurately or not). Such semantic assumptions accompany two other sets of axioms in an axiomatized scientific theory (Bunge, 1967d, 1967f, 1973a, 1973b). One of the sets is constituted by the basic factual formulas of the theory, such as the hypotheses concerning the constitution, structure, and mode of change of the systems concerned. The other set of basic elements is formed by the assumptions concerning the mathematical nature of the concept occurring in the previous statements. In other words, the axiom base A of a scientific theory can be partitioned into the following sets:

M = *The mathematical postulates* (e.g. function differentiability)

F = *The basic factual formulas* (e.g. law statements)

S = *The semantic formulas* (designation and denotation rules and representation assumptions).

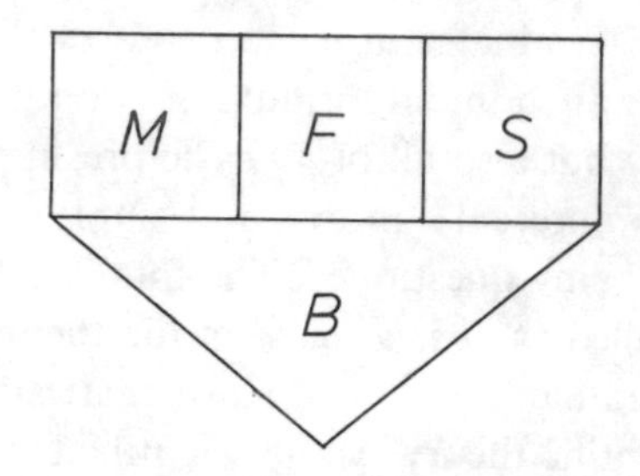

Fig. 5.1. The foundations of a scientific theory: the background B, the mathematical assumptions M, the factual hypotheses F, and the semantic formulas S.

Prima facie F is purely factual whereas M and S are nonfactual. However, a formula has a factual purport provided it is "read in the light" of both the mathematical postulates and the semantic assumptions: otherwise the formula is mathematically as well as factually senseless. Hence it is impossible to single out a subset of A as constituting the factual gist of T. Put in a positive way: All of the nonlogical axioms of a theory constitute its specific factual gist. To repeat with solemnity:

DEFINITION 5.8 Let T be a theory with axiom base $A = M \cup F \cup S$, where M is the set of mathematical axioms, F the specific factual assumptions, and S the semantic formulas of T. Then

(i) the *specific mathematical gist* of T equals $M \cup F$;

(ii) the *specific factual gist* of T equals $M \cup F \cup S$, i.e. the specific extralogical gist of T – in short the whole of A.

This is no concession to semantic holism. We are not claiming that the factual purport of a theory is unanalyzable, but that it is the outcome of combining three distinct reactants: mathematical postulates, factual hypotheses, and semantic formulas. We can distinguish these ingredients but not separate them: M is idle without F, F meaningless without both M and S, and S pointless without F. The three components are no less united and no less distinguishable than the sides of a triangle.

2.4. *Changes in Gist*

The 1960's heard a number of spirited quarrels concerning the changes in meaning accompanying scientific revolutions. Unfortunately there was more noise than light about them, for they were not preceded by an elucidation of what was rightly deemed to change, namely meaning. Let us see whether we can do any better with the help of the tools wrought in Sec. 2.3. (A fuller investigation of the question will have to wait until Ch. 7, Sec. 3.3.)

Our Definitions 7 and 8 provide a simple way of elucidating the concept of change in one of the components of meaning, namely basic purport or gist. Let us see how this works in a simple case before proceeding to a general definition.

The simplest algebraic structure is a set S together with a binary operation $\circ$ in S. This structure, called a *groupoid*, may be characterized by a single axiom;

$$A = \ulcorner \circ \text{ is a binary operation on the set } S. \urcorner$$

Let us now add the condition that $\circ$ be associative. The ensuing structure is called a *semigroup*. The axioms of the theory of semigroups are the previous statement A and

$$B = \ulcorner \circ \text{ is associative in } S. \urcorner$$

Consequently by Definition 7 the specific extralogical gist of the theory of semigroups is $\{A, B\}$. The only difference between this theory and the theory of groupoids is constituted by the additional postulate B. That is

$$\delta\,(\text{Semigroup theory, Groupoid theory}) = \{A, B\} - \{A\} = \{B\}.$$

The generalization suggests itself: The difference in basic purport (gist) between two theories consists of the axioms they do not share. More explicitly we propose

DEFINITION 5.9 Let T and T' be two nonlogical theories with axiom bases A and A' respectively. Then the *difference in specific extralogical gist* between T and T' equals the symmetric difference between A and A':

$$\delta(T, T') = A \Delta A' \equiv A - A' \cup A' - A.$$

It sometimes happens that two scientific theories share a mathematical formalism but interpret it differently: that is, they have the same M's and F's but different S's. At times this semantic difference is due to a difference in the referents, at other times the referents are the same but the predicates are assumed to represent different traits of their referents. An example of the first kind of semantic unlikeness is afforded by the theories of contagion and of the spread of rumours. An example of the second kind of semantic difference is the variety of rival interpretations of quantum mechanics. In either case the difference lies in the semantic formulas of the theory. More exactly we have

DEFINITION 5.10 Let T and T' be two theories with the same mathematical gist but different sets S and S' respectively of semantic formulas. Then the *difference in specific factual gist* between T and T' equals the symmetric difference between S and S':

$$\delta(T, T') = S \Delta S'.$$

From a semantic point of view it does not matter whether T and T' are coexistent or successive. Nor is it indispensable, in order ot use the concept of change in gist, to axiomatize the theories of interest. Axiomatization is needed only to determine the exact amount of change in basic purport or gist.

The problem of changes in both full sense and reference, i.e. the question of meaning change, will be taken up in Ch. 7, Sec. 3.3. We must now turn to the dual of purport, namely import, which is the other component of sense. (Since constructs deriving from the same source have the same purport, they must differ in import if they are to differ at all.)

3. Sense as Import or Logical Progeny

3.1. *The Logical Progeny of a Construct*

In Sec. 1 we characterized the import of a construct as its logical offspring. Since the concept of a filter will allow us to elucidate the notion of logical progeny, we may as well take a look at filters. We can do it quickly because a filter is the dual of an ideal – a character we met in Sec. 1.2.

In Sec. 1.1 we learned that every closed set of constructs is a Boolean algebra B and, *a fortiori*, a lattice. When tracing down the followers of an element (predicate or statement) in such an algebra we wish to avoid the surprise of ending up in the null element, for this spells disaster just as the unit element spells triviality. To get rid of $\square$ we take a certain subset of B known as a proper filter. But first the general notion.

A filter in a lattice L is that subset F of L which contains all the followers of any given element of L as well as the meets of any two elements of F. More precisely,

If L is a lattice, then F is a *filter* in L iff F is a nonempty subset of L satisfying the following conditions:

$F1$ For all $x \in F$ and $y \in L$, if $x \leqslant y$ then $y \in F$.

$F2$ For all $x, y \in F$, $x \wedge y \in F$.

If the null element is missing from F the filter is called *proper*. Since our construct lattices are Boolean, our filters are more than proper: they are ultrafilters. In other words, a proper filter F in a Boolean algebra B is called an *ultrafilter* just in case, for each $x \in B$, either x or its complement $\bar{x}$ is in F but not both.

Our closed sets of constructs are ultrafilters provided we remove the null construct from them. This will suffice to guarantee that they are consistent. For, if x is in F, then $\bar{x}$ will be automatically excluded from F, as a consequence of which $x \wedge \bar{x}$ will be out as well. In short, we have the following

LEMMA A closed set of constructs is *consistent* iff it is an ultrafilter.

Finally, a filter can be analyzed into partial filters (subfilters). Every member of the base lattice generates its own family, that constitutes the private filter of that element. Formally:

For every element x in a lattice L, the set $\{y \in L \mid x \leqslant y\}$ is called the *principal filter* generated by x. Symbol: $)x(_L$.

Since in the case of constructs the ordering relation is that of entailment, for our purposes we need

DEFINITION 5.11 Let x be in a closed set $\mathbb{C}$ of constructs. Then the *logical offspring* of x in $\mathbb{C}$ is the principal filter generated by x in $\mathbb{C}$:

$$)x(_{\mathbb{C}} = \{y \in \mathbb{C} \mid x \vdash y\}.$$

The offspring or progeny of a nontautologous construct depends on the context in which it occurs. (Tautologies, being deductively barren, have no progeny.)

It should come as no surprise the manner in which these concepts, taken from the algebra of logic, will presently be exploited for semantic purposes.

3.2. *Import*

We are now in a position to elucidate the idea that the import of a construct is the collection of constructs it generates or "contains":

DEFINITION 5.12 The *import* of a construct x in a closed set of constructs $\mathbb{C}$ equals the former's logical offspring, i.e. the principal filter it generates:

$$\mathscr{I}mp_{\mathbb{C}}(x) =)x(_{\mathbb{C}} = \{y \in \mathbb{C} \mid x \vdash y\}.$$

Example 1 The import of a tautology is nil. *Example 2* Take identity to be defined by Leibniz' law. Then the import of "=" includes the reflexivity, symmetry and transitivity of "=". (This would have puzzled Austin, who regarded "same" as being devoid of positive meaning. But this is excusable: neither Wittgenstein nor his followers had a semantic theory of meaning.) *Example 3* The import of the law of the vibrating string includes the superposition principle (⌜The superposition of any two oscillations is a third oscillation.⌝)

If $\mathbb{C}$ happens to be a *consistent* closed set of constructs, i.e. one in which no contradictions occur, then the import of a construct x in $\mathbb{C}$ will be the principal *ultra*filter generated by x in $\mathbb{C}$. Because filters are the duals of ideals, and ultrafilters the duals of maximal ideals, we see that import and purport are the dual of one another, hence mutually complementary rather than rivals. (More on this in Sec. 4.)

It would be pleasant, but it seems chimerical, to obtain general relations among the principal filters generated by arbitrary constructs in a closed set of constructs, so as to compute, say, the import of the conjunction of two statements in terms of their individual imports. A similar obstacle was met in Sec. 2.1 with reference to principal ideals. A definite algebraic structure is found only in each offspring of a given construct as well as in the total context. The only simple statement we can make about the relations in import between two constructs is the dual of Theorem 4, namely

THEOREM 5.5 For any two constructs x, y in a closed set of constructs $\mathbb{C}$,

(i) $\mathscr{I}mp_{\mathbb{C}}(x)=\mathscr{I}mp_{\mathbb{C}}(y)$ iff $x=y$;

(ii) If $x \vdash y$ then $\mathscr{I}mp_{\mathbb{C}}(x) \supseteq \mathscr{I}mp_{\mathbb{C}}(y)$

On the other hand it is possible and advantageous to extend the concept of import to both a set of constructs and their conjunctions. Let $\mathbb{C}_0$ be a subset of a closed set of constructs $\mathbb{C}$. Usually there will be a number of filters containing $\mathbb{C}_0$. And the intersection of any nonempty set of filters is a filter provided it is nonempty. It is, of course, the smallest filter containing $\mathbb{C}_0$ and for this reason it is called the *least filter generated* by $\mathbb{C}_0$. Accordingly we stipulate

DEFINITION 5.13 Let $\mathbb{C}_0$ be a subset of a closed set of constructs $\mathbb{C}$. Then the *minimal import* of $\mathbb{C}_0$ equals the least filter generated by $\mathbb{C}_0$ in $\mathbb{C}$.

Example Let $\mathbb{C}$ be a theory and $\mathbb{C}_0$ some of its axioms. Then $\mathscr{I}mp_{\mathbb{C}}(\mathbb{C}_0)$ is the smallest collection of theorems (including $\mathbb{C}_0$) entailed by $\mathbb{C}_0$. Look at the picture:

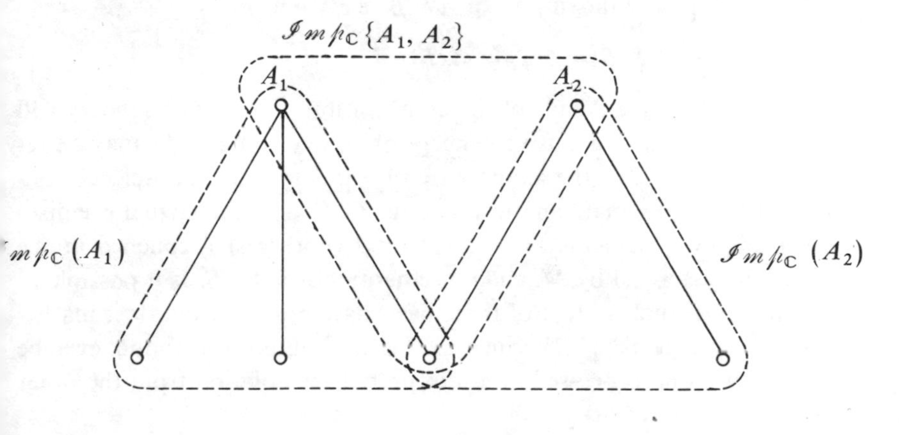

We shall get a far more interesting semantic concept if we consider the offspring of a bunch of axioms taken jointly. Let us look into this.

3.3. *Theory Content*

Consider a scientific theory T based on a set $\{A_1, A_2,\ldots, A_m\}$ of specific axioms as well as on a background $\{B_1, B_2,\ldots, B_n\}$ of presuppositions. The latter include, among others, all the logical and mathematical theories underlying T. Let us conjoin all such initial assumptions and examine their import. Call

$$A = A_1 \,\&\, A_2 \,\&\ldots\&\, A_m,$$

The import of A & B is, of course, the offspring of A & B. One part of this import is purely logical: it consists of the set L of tautologies included in B or entailed by A & B. If we remove this part L we are left with the extralogical import, or *content*, of the axiom basis A & B. More explicitly, we make $B = B_1 \,\&\, B_2 \,\&\ldots\&\, B_n$.

DEFINITION 5.14 Let T be a theory based on a conjunction A of axioms and another conjunction B of presuppositions, and call L the set of tautologies in B or in the offspring of A & B. Then

(i) the *import* of the axiom base A & B equals the principal filter generated by A & B, i.e.

$$\mathscr{I}mp(A \,\&\, B) =$$

(ii) the *content*, or *extralogical import*, of the axiom base A & B equals the complement of the import of A & B relative to L:

$$\mathscr{C}(A \,\&\, B) = \mathscr{I}mp(A \,\&\, B) - L.$$

Going into some details of the composition of a scientific theory will allow us to grasp the elusive concept of theory content. As may be recalled from Sec. 2.3, the axiom base of a theory in factual science consists of (*a*) purely mathematical axioms M, (*b*) specific factual premises (such as law statements and subsidiary hypotheses) F couched in the terms characterized by M, and (*c*) semantic formulas S. Is it possible to isolate the factual content of these basic assumptions from their mathematical ingredients? No, because the factual components cannot even be stated, let alone processed and interpreted, in isolation from the other two. What can be neatly detached is the mathematical content of the initial assumptions: this is done by simply removing the semantic axioms S and finding the progeny of the remaining premises. All this is encapsulated in

DEFINITION 5.15 Let T be a theory with background B, mathematical assumptions M, factual assumptions F, and semantic formulas S, where B, M, F and S are taken jointly. Then

(i) the *mathematical content* of the axiom base A & B, or *formalism* of the theory T, equals the offspring of B & M & F with exclusion of L, i.e.

$$\mathscr{C}_M(A \,\&\, B) = \mathscr{I}mp(B \,\&\, M \,\&\, F) - L;$$

(ii) the *factual import* of the axiom base, or *factual content* of the theory T, equals the offspring of B & M & F & S with exclusion of L, i.e.

$$\mathscr{C}_F(A \,\&\, B) = \mathscr{I}mp(B \,\&\, M \,\&\, F \,\&\, S) - L.$$

That is, the factual content of a scientific theory coincides with the factual import of its axiom base, which is in turn identical to its extralogical import. In other words, while scientific theories have detachable, hence portable mathematical formalisms, their factual content is not detachable from the formalism. (*A fortiori* they have no purely empirical or observational content.) The extralogical import of the axiom base of a theory in factual science is both mathematical and factual.

Note that the notion of content, though dependent upon the metalogical concept of deducibility, is independent of the semantic concept of truth of fact. Hence a theory with a rich content may well prove to be false whereas a more modest theory may turn out to be approximately true. Consequently, *pace* Popper (1966), the degree of truth of a theory cannot be measured by its content. If it were so measurable then no actual empirical tests for truth would be called for: a semantic analysis of a theory would determine its truth value.

We close by noting a possible application of the foregoing ideas to model theory. Intuitively the richer an abstract theory ("formalized language") the fewer models it is bound to have. Thus there are less examples of a semigroup with unit element than of a semigroup. This suggests introducing a comparative concept of theory versatility, or its dual, rigidity, namely thus:

DEFINITION 5.16 Let T and T' be two abstract theories with comparable mathematical contents $\mathscr{C}(T)$ and $\mathscr{C}(T')$ respectively. Then T is *more versatile* (or *less rigid*) than T' iff $\mathscr{C}(T) \subseteq \mathscr{C}(T')$.

3.4. *Empirical and Factual Content*

The various philosophies of science may be regarded as so many views on the content of scientific theories – i.e. as so many semantics of science. Here goes, in telegraphic fashion, a sample of the most interesting among those views:

(1) *Conventionalism.* Scientific theories are handy tools for the processing of experience (or the systematization of appearances). They have neither a factual nor an empirical content: $\mathscr{C}_F(T) = \mathscr{C}_E(T) = \emptyset$.

(2) *Positivism.* Scientific theories are economic summaries and articulations of data. They have an empirical (observational, operational) content only: $\mathscr{C}(T) = \mathscr{C}_E(T)$.

(3) *Mellowed positivism (standard view).* Scientific theories are systematizations of experience propped up by heuristic devices, i.e. the so-called theoretical terms, which have no representational function. Part of the content (our "import") of a statement s in a scientific theory is observational (empirical, phenomenal) provided s is linked to observational items via correspondence rules: $\mathscr{C}_E(s) \subseteq \mathscr{C}(s)$.

(4) *Empirical realism.* Theoretical constructs may refer to unobservables even though their meaning is given by the connections established within the theory between the constructs and observational statements built with the observational vocabulary of the theory. Every scientific statement s has both an empirical content and a factual one (its "surplus meaning"): $\mathscr{C}(s) = \mathscr{C}_E(s) \cup \mathscr{C}_F(s)$.

Of these four views the closest to our own is the last, according to which scientific hypotheses have "a surplus meaning over against their evidential basis" (Feigl, 1950). (For elaborations see Rozeboom (1962, 1970). For criticisms see Hempel (1950) and Nagel (1950).) Although timid realism is preferable to the previous forms of empiricism, it is not completely satisfactory: keeping the myth of the observational vocabulary as belonging to every theory and as the source of meaning of the theoretical vocabulary does violence to scientific theories, all the concepts of which are theoretical. Furthermore it has the fatal consequence that it becomes impossible to ascertain what exactly the theory is about: reference becomes indeterminate (Rozeboom, 1962, 1970). Rather than tolerate such reference uncertainty, as Rozeboom advocates, we regard it as sufficient for disqualifying the theory: reference indeterminacy makes not only for meaning indeterminacy but renders empirical tests impossible. We must either retreat to undiluted empiricism or go realists all the way. Since the former has proved a failure we shall try the latter.

We claim that no scientific theory could possibly embrace its own evidence, and that only theories concerning sentient organisms may be said to have an empirical content. The reasons for denying that the evidence relevant to a theory is part of its content have been given in detail elsewhere (Bunge, 1967a, 1969, 1973b). They will be summarized here. Firstly, the computing of a prediction with the help of a theory requires additional theoretical assumptions (e.g. on the particular composition or structure of the referent) as well as empirical data – and neither can be found in a theory unless it happens to be an *ad hoc* systematization of data. (If the special assumptions were already included in the theory then the latter would be inapplicable to further problems. And if the data were part of the theory then observations would be redundant.) Secondly, the production of empirical data that could serve as evidence for or against a theory calls for the cooperation of auxiliary theories employed in the design and interpretation of the empirical operations – i.e. the evidence could not have been predicted on the strength of the substantive theory alone. Should a scientific theory have an empirical content, then this content would not consist in any evidence but in statements concerning some sector of human experience, such as learning or working, considered as an external item. (More on this in a while.) These are some of the reasons for adopting a realist semantics.

Of the three other views sketched above, conventionalism is untenable if only because it renders empirical testing, as well as the quest for truer and deeper theories, utterly pointless. The second view, undiluted positivism, is still popular in scientific circles but it has been discredited by philosophers long ago, so it need not detain us. (Which does not mean disparaging catechisms for the conversion of the infidel.) Only the third view, which we have dubbed *mellowed positivism*, is still alive and influential though on the decline. It is also the only view that has been expounded and examined in detail (Carnap, 1956, 1963a, 1963b, 1966; Hempel, 1958; Braithwaite, 1959, 1962; Wojcicki, 1966; Przelecki, 1969;

etc.). The way this doctrine handles the concept of empirical or observational content is roughly as follows.

The standard view assumes, first of all, that every scientific theory contains not only theoretical terms, such as 'temperature', but also observational or phenomenal ones such as 'hot'. Second, only the theoretical terms are assumed to be subject to the theoretical statements Θ of the theory. They are interpreted, albeit only in part, by reference to the observational terms: this interpretation is performed by the so called "correspondence rules" C of the theory. Third, the content of a statement is, roughly, what we have called its import. And the observational (or empirical) content of a statement s is defined as the class of all non-logically true observational statements implied by s – i.e. the set of all those consequences of s that contain observational terms but no theoretical ones. In particular, the observational content of s relative to a theory composed of theoretical statements Θ and correspondence rules C (both sets taken conjunctively), is the observational content of $s \& \Theta \& C$. (For a brief and lucid exposé see Carnap (1963b). For details cf. Suppe (1971). For the alleged definability of theoretical terms as functions of empirical ones: Carnap (1961, 1966).)

The root trouble with this view is, of course, that it rests on the unquestioned assumption that every scientific theory does contain observational or phenomenal terms such as 'hot', 'blue', and 'rough'. In point of fact theories contain only theoretical concepts and these are not interpreted with reference to qualia but to supposedly real things (e.g. genes) and their properties (e.g. the mutation frequency of a given gene). The semantic assumptions in a well constructed theory are not concept-concept affairs in which one of the two partners concerns an experiential item: they are concept-fact correspondences (Ch. 3, Sec. 4.1). Now, if there are no purely observational (hence unanalyzed) concepts in a scientific theory, then there is no observational (or phenomenal or operational or empirical) content left. And since (unlike Feigl) Carnap, Hempel and Braithwaite make no room for factual reference, no content at all is left.

This unintended disaster could have been averted if any real scientific theories had been analyzed in the light of the standard semantics, and if the quest for the Holy Grail of the empirical "basis" of science had been given up in time. If theories are regarded instead as partial and symbolic representations of reality, then a far simpler and realistic semantics of science emerges. Every concept in a theory is pronounced *theoretical* if it is elucidated by the theory or borrowed from the background of the theory. And if the concept has a factual reference then any statement that contains the concept has a factual sense. The facts referred to, or represented by, the theoretical statement are not those the empirical evidence is concerned with. (In fact an evidence statement is restricted to observational items, often under control with the help of instruments describable with the assistance of a number of theories.) No scientific theory is concerned with its own test, much less with the mental states of the experimenter. Finally, the factual content of a statement in a scientific theory is determined by all three basic components of the theory: M, F, and S. (Recall the preceding subsection.)

According to our view, then, all scientific theories have a factual content and none of them embraces its own evidential or empirical "basis". However, a scientific theory *can* have an empirical content – provided it *is* about experience. This is the case with most, perhaps all, theories in the sciences of man. Like all other scientific theories, those in psychology, sociology and history concern facts and so have a factual content (even if inaccurate). But (*a*) these facts are treated as external objects epistemologically on a par with physical facts – which is necessary for objectivity, and (*b*) those theories contain no reference to their own empirical tests – which alone renders tests significant. For example, a model of the decay of empires may refer to feelings and motives – of the imperial subjects not the historian's; and its evidence may consist in documents and material testimonials unearthed with advanced physical techniques – but the model won't be concerned with such means for its own validation or invalidation.

In summary, if s is a theoretical statement then the factual content of s equals its extralogical import. Should s belong in natural science, its empirical content will be nil. But if s belongs in the sciences of man, then s will have an *empirical content* included in its factual content and independent of any evidence relevant to s. No theory has an observational or experiential or phenomenal content: experience is lived through (*erlebt*) not beaten into theory. This view may be dubbed *realist semantics.* It boils down to shifting the focus from experience (always subjective) to fact, and consequently prevents us from falling into subjectivism. It suggests that we stop chasing the ghost of the alleged observational content of theories as well as its attendant ghosts – Ramsey sentences and Craig's reduction.

Now that operationism lies buried we can afford to write a nice epitaph on it. The deceased, though mistaken in the means it employed, was sincere in pursuing a worthy goal – namely the uprooting of inscrutables and the testability of hypotheses. This explains the large following it had and still has in the scientific community. Such a noble cause can now be defended in the following way: Although theoretical statements may lack an empirical import, they should acquire one when conjoined with suitable empirical statements relevant to the former. How this strategy can be implemented may be exemplified as follows.

Consider the statement schema

$$\ulcorner\text{There is a gravitational field in region } x.\urcorner \tag{1}$$

belonging to some theory of gravitation. Although fields defy direct observation, according to operationism (1) *means* that, if somebody places a test body y in region x, he will observe y to move (cf. Carnap, 1936). But we know this to be a gross misinterpretation of (1), which refers only to a field, not to a test body, let alone to an observer. However, our construal of the notion of sense as import suggests the following repair: conjoin (1) with

> ⌜A body acted on by a gravitational field is accelerated & y is an observable body & y is at x [whether or not it was put there by somebody] & z is suitably equipped to observe y & z observes y.⌝ (2)

This new statement schema has one theoretical component, viz., the first conjunct, which is a law statement. The remaining components of (2) are empirical and moreover relevant to both the theoretical component and the original formula (1). As a consequence of this empirical ingredient, the conjunction (1) & (2) will entail observational statements and consequently will have an empirical import. Part of the latter will be

$$\ulcorner z \text{ sees } y \text{ moving}\urcorner, \tag{3}$$

which is in fact entailed by (1) & (2). In short, although the "operationally meaningful" statement (3) is not in the import of the original theoretical statement (1), it is in the import of the conjunction of the latter with further items, some of which are empirical. What holds for a single statement is true, *a fortiori*, of entire theories. That is, any substantive scientific theory can be enriched with further statements, at least some of them empirical, to yield observational consequences and thus get ready for observational tests. Thus the goal of operationism, namely to ensure testability, is saved without tergiversating the nature of scientific theories. *Requiescat in pace.*

3.5. *Changes in Import and Content*

The content of a theory is determined by its axiom base but is never known in its entirety. The content of a theory is fixed, however, once its axioms have been adopted. Moreover the content is infinite because so is the set of logical consequences of the axioms. And if the theory involves continua, then its content is nondenumerably infinite: the axioms generate an uncountable offspring. On the other hand the known part of the total content of a theory is of course finite and keeps growing as long as the theory stays alive – i.e. as more theorems in it are proved and fresh applications of it are made. This is the case not only with new theories but also with old theories that are still going strong, such as classical mechanics. There is growth in both cases although the growth *tempi* and mechanisms are not the same.

The growth of knowledge and its mechanisms, internal and external, is a fascinating subject but it lies outside the scope of semantics: they are the concern of the philosophy, history, psychology, and sociology of science. What the semanticist can do in this respect is to compare the import of alternative axiom bases, whether successive or contemporaneous. He can thus help other scholars by supplying measures of the net differences in import or in content, differences that may be interpreted as accompanying the replacement of one theory by another in the favor of the scientific community. Definition 15 will help us to contrive such a measure of change in scientific theories.

According to Definition 15, the factual content of a scientific theory equals the sum total of its nonlogical formulas. Hence the difference in content between two theories boils down to whatever nonlogical formulas the theories faill to share. More precisely, we introduce

DEFINITION 5.17 Let T and T' be two scientific theories with backgrounds B and B', mathematical assumptions M and M', factual assumptions F and F', and semantic assumptions S and S' respectively. Then the *difference in factual content* between T and T' equals the symmetric difference of their factual contents:

$$\delta(T, T') = \mathscr{I}mp(B \,\&\, M \,\&\, F \,\&\, S) \,\Delta\, \mathscr{I}mp(B' \,\&\, M' \,\&\, F' \,\&\, S') - L.$$

Since T and T' are infinite sets, their difference may prove infinite as well. On the other hand the gist differences introduced in Sec. 2.4 (Definitions 9 and 10) look more manageable because they concern only the axiom bases. However, this greater simplicity of the difference in gist does not render it more suitable as a measure of the difference in sense. In fact, since a theory can usually be axiomatized in alternative ways, the gists of two different axiomatizations of a given body of knowledge are bound to differ even though the total content of the theory is unaltered by the different axiomatic arrangement. Obviously we need both measures. And none of them is needed for practical purposes, such as rewarding the scientist who manages to produce the richest of two rival theories. The assumptions, definitions and theorems in our semantics serve only the modest purpose of elucidating some interesting metascientific ideas. (We shall come back to meaning change in Ch. 7.)

4. FULL SENSE

In the previous chapter we studied the content or interior of a construct apart from its exterior or extension and regardless of its place in a body of knowledge. The corresponding concept of sense, i.e. intension, is the one that suits constructs which, without being totally stray, belong to no definite system or closed context.

In the present chapter we have focused on systemic constructs and have distinguished two components of their sense – namely their purport and their import. The purport and the import of a construct are certainly not external to it but they are not strictly intrinsic either: they are essentially context-dependent. A closed context $\mathbb{C}$ may be pictured as a tree oriented from top to bottom. An arbitrary node or branching point x of $\mathbb{C}$, i.e. a systemic predicate or proposition, is assigned two basic sense components. One is its purport or upward sense $\mathscr{P}ur_{\mathbb{C}}(x)$, which equals the principal ideal generated by x in $\mathbb{C}$. The other component is the import or downward sense $\mathcal{I}mp_{\mathbb{C}}(x)$, equal to the principal filter generated by x in $\mathbb{C}$. See Figure 5.2.

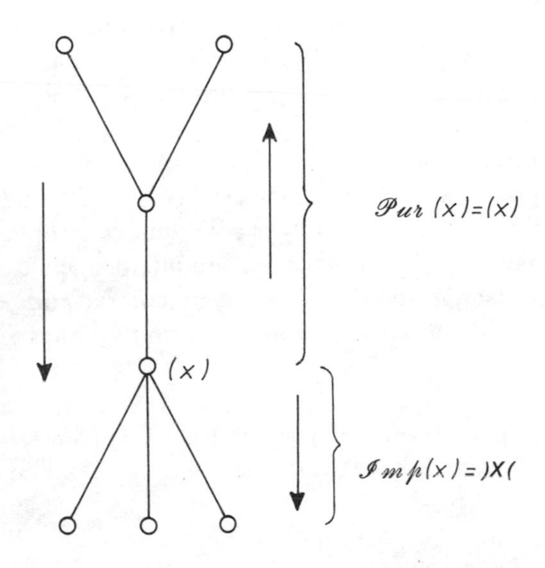

Fig. 5.2. A fragment of a conceptual system. The two components of the sense of a construct x in the system: the purport $\mathcal{P}ur(x)$ and the import

The sense of a construct in a closed context has no components other than its purport and its import. Since both are sets, or may be taken as such, we can build their union. It seems natural then to regard this union as the full sense of the construct of interest. More precisely, we offer

DEFINITION 5.18 Let x be a construct in a closed set of constructs $\mathbb{C}$. Then the *full sense* of x in $\mathbb{C}$ is the union of the purport and the import of x in $\mathbb{C}$:

$$\mathcal{S}_{\mathbb{C}}(x) = \mathcal{P}ur_{\mathbb{C}}(x) \cup \mathcal{I}mp_{\mathbb{C}}(x) = (x)_{\mathbb{C}} \cup)x(_{\mathbb{C}}.$$

This decomposition of the total sense of a construct is consonant with the scientific practice of regarding meanings as being either assumed or derived. Thus if it is *assumed* that the values of a certain function X of time represent positions (or concentrations or population densities), then it can be *inferred* that the value of its time derivative $\dot{X}$ at $t=0$ represents the initial velocity (or rate of change of concentration or population density). For this reason only the factual primitives of a theory need be assigned a factual sense by semantic formulas: the derived (defined or proved) constructs get their meaning from the source, i.e. the axiom base. In other words, in a scientific theory *sense flows downwards*, from assumption to consequence – not upwards, as the positivist doctrine has it.

Since a tautology follows from any proposition and entails all of the other tautologies in a given logic, the full sense of a tautological construct in a context equals the totality of propositions in it plus the underlying logic. Likewise all contradictions are interdeducible, so that the sense of any of them includes all; and since a contradiction entails anything, its sense is the totality of propositions in its context. That is, we have proved

COROLLARY 5.2 Let t be an analytic construct occurring in the logic L underlying a context $\mathbb{C}=\langle S, \mathbb{P}, D\rangle$. Then
(i) $\mathscr{S}_{\mathbb{C}}(t)=S\cup L$;
(ii) $\mathscr{S}_{\mathbb{C}}(\neg t)=S$.

In a strictly logical context, i.e. for $S=L$, the sense of a tautology equals logic itself. In other words, logical truths "say" nothing when confined to logical theories and they "say" too much – in fact everything – when associated with extralogical bodies of knowledge.

Because an arbitrary proposition entails all of the tautologies, it falls between the above extremes of minimal and maximal sense:

COROLLARY 5.3 Let p be a statement occurring in a context $\mathbb{C}=\langle S, \mathbb{P}, D\rangle$ with underlying logic L. Then

$$L\subseteq\mathscr{S}_{\mathbb{C}}(p)\subseteq S\cup L.$$

To obtain the extralogical (in particular factual) sense of a proposition we must then subtract L from its full sense:

DEFINITION 5.19 Let p be a proposition in a context $\mathbb{C}=\langle S, \mathbb{P}, D\rangle$ with underlying logic L. Then the *extralogical sense* of p in $\mathbb{C}$ is

$$\mathscr{S}_{\mathbb{C}\bar{L}}(p)=\mathscr{S}_{\mathbb{C}}(p)-L.$$

If we know the full senses of two constructs we can ascertain whether they are comparable and, if so, which contains which. Thus the sense of "river" is included in that of "mountain river": the latter is the richer of the two. In general we have

DEFINITION 5.20 Of two constructs with comparable senses, the *richer* (or *more complex*) is the one including the other:
If c and c' are in $\mathbb{C}$, then c is *richer* than $c' =_{\text{df}} \quad \mathscr{S}_{\mathbb{C}}(c) \supseteq \mathscr{S}_{\mathbb{C}}(c')$.

The richer of two concepts is also the more specific and complex, i.e. the less generic and less simple. The last convention serves then as a definition of *semantic complexity* – or its dual, semantic simplicity. But of course even if two constructs are not comparable in sense they may share a grain of sense. This notion of sense overlapping is exactified by

DEFINITION 5.21 The *core sense* of a set $\{c_i \in \mathbb{C} \mid 1 \leqslant i \leqslant n\}$ of constructs equals their shared sense:

$$\mathscr{S}_{core}\{c_i \in \mathbb{C} \mid 1 \leqslant i \leqslant n\} =_{df} \bigcap_{i=1}^{n} S(c_i)$$

Example Each of the different explicata e_i, for $1 \leqslant i \leqslant n$, of a given coarse or intuitive concept c, is supposed to recapture a part of the sense of the latter:

$$\mathscr{S}(c) \supseteq \mathscr{S}_{core}\{e_i \in \mathbb{C} \mid 1 \leqslant i \leqslant n \;\&\; e_i \text{ explicates } c\}.$$

If we focus on the specific contribution a basic construct makes, and brush aside the background of the theory, i.e. if the search for ancestors is stopped at the basic construct a itself, the purport of a will include just a, which is already included in its principal filter or import $\mathscr{I}mp_T(a)$. Hence we have

COROLLARY 5.4 The full specific sense of a basic construct a, characteristic of an axiomatized theory T, equals the import of a, i.e.

$$\mathscr{S}_T(a) - \mathscr{I}mp_T(a).$$

Hence the import of a construct measures its importance. On the other hand the greater the purport of a construct, i.e. the larger its dependence upon other constructs, the smaller its importance.

Because in a closed context $\mathbb{C}$ there is no source of sense other than $\mathbb{C}$ itself, intension has no separate existence in it: it is part of the full sense. More explicitly, we conjecture

PROPOSITION 5.1 The intension of a construct x in a closed set of constructs $\mathbb{C}$ is included in the full sense of x:

$$\text{If } x \in \mathbb{C}, \quad \text{then} \quad \mathscr{I}(x) \subseteq \mathscr{S}_{\mathbb{C}}(x).$$

Example Consider the predicates "triangular" ($\triangle$), "equiangular" (A), and "equilateral" (L), all of them referred to closed plane figures in Euclidean geometry (E). Since $A \neq L$, $\mathscr{I}(\triangle \,\&\, A) \neq \mathscr{I}(\triangle \,\&\, L)$. But, because triangle equiangularity implies triangle equilaterality and conversely, $\mathscr{S}_E(\triangle \,\&\, A) = \mathscr{S}_E(\triangle \,\&\, L)$. Moreover the full sense of either complex predicate includes not just A and L but also the whole lot of derived properties of equilateral triangles, which their intensions do not.

From the preceding it follows that, if a construct happens to be basic (i.e. undefined or assumed) in a given context, then its intension coincides with its full sense and also with its import:

PROPOSITION 5.2 The intension of a basic construct a in a closed set $\mathbb{C}$ of constructs equals the full sense of a:

$$\text{If } a \in \mathbb{C} \text{ is basic in } \mathbb{C}, \quad \text{then} \quad \mathscr{I}(a) = \mathscr{S}_{\mathbb{C}}(a) = \mathscr{I}mp_{\mathbb{C}}(a)$$

A change in the basic principles that determine the sense of a construct may alter the latter even if we keep calling it by the same name. As Hilbert explained to Frege (who disliked axiomatic definitions), "Each axiom contributes something to the definition [of the concept of interest] and, as new axioms are introduced, the concept is accordingly changed" (Hilbert, 1900). In short, meaning is contextual.

This concludes the study of the concepts of sense we started in Ch. 4. It may well happen that our exactifications fail to capture the reader's intuitions. One man's sense is another's nonsense. This is quite in order: no technical concept is supposed to satisfy everybody's presystematic ideas. (Recall the example illustrating Proposition 5.1.) If the proposed elucidations are found unacceptable for technical reasons, then surely better theories can be contrived. But once a theoretical explication is accepted, be it in science or in philosophy, it is one's intuitions that are expected to adjust to the theoretical concept – until a better exactification becomes available. (Cf. Bunge, 1962.)

5. Conclusion

There are two main views on sense. One is extensionalism, according to which constructs have an exterior, or extension, but no interior or sense. We have found this, the hollow construct view, unacceptable not only in relation to factual science but also to pure mathematics (Ch. 4, Sec. 1.2). The opposite view regards all constructs other than sets as having a content. But there are many differences of opinion concerning the nature of this content (Ch. 4, Sec. 1.1). These differences are manifested, in particular in the question of the relations between the sense of a whole and the sense of its parts. There are two extreme views on this matter: semantic atomism and semantic holism. Let us take a glimpse at them and locate our own view in this perspective.

Semantic atomism holds that "The sense of the whole is determined by the senses of its parts". It has been the most popular view among illuminist philosophers from Hobbes right down to Montague (1970) and Scott (1970). But no semantic atomist seems to have constructed a suitable theory containing formulas that determine the sense of a construct as a function of the sense of its components. In short, semantic atomism has remained at a programmatic stage. The opposite extreme, semantic holism, maintains that "The sense of every item is determined by the whole of knowledge". Its most eminent proponent in recent times has been Quine (1952). This view too is programmatic and moreover doomed to stay in this stage, because nobody can handle the sum total of human knowledge.

Our own semantic theory takes a *via media* and therefore cannot be as charming as either of the extremes. Indeed, on the one hand we have proposed definite formulas for the intension of a composite construct in terms of the intensions of its components (Ch. 4, Sec. 2). However, this analysis is not a term by term resolution and, moreover, it offers no means for determining the sense of a component. If asked how the latter is to be determined, we offer still another theory, according to which the sense of a construct can be determined provided the construct occupies a definite place in a conceptual system. Hence, to the extent to which our theory provides for the sense analysis of a whole, it implements a modest version of the atomistic program. And since the full sense of a part is shown to depend on its role in a whole, our theory satisfies the holistic injunction not to isolate. In short our semantics constitutes a merger of the best (others would say the worse) of both extremes. If accepted, it becomes understandable why each of the two extremes has its appeal – and why none of them got beyond the program stage.

At first blush our *via media*, by admitting that the sense of a (theoretical) construct is relative to the theory in which it occurs, and is thus liable to become a different construct if inserted in a different theory, would seem to encourage subjectivism. Perish the thought. Firstly, only the sense of a theoretical or systemic construct is determined by the whole theory in which it occurs. Extrasystematic constructs, such as the common observational statements, are not theory-dependent – but then their sense is uncertain. Secondly, it is not that "the meaning of a scientific construct is theory-laden", as the fashionable trend has it, but the other way around: theories (even logical theories) are meaning-laden. The meaning of any component of a theory is relative to the theory, i.e. it depends on the role it plays in the theory, but is not "theory-laden" – whatever this metaphor may mean. Thirdly, every authentic scientific theory, even if false, is objective – ergo so is every factual component of it. It is objective both semantically (by its reference to entities that are supposedly out there) and methodologically – for being publicly scrutable. On the other hand the attempts to defend the objectivity of science by anchoring it to an experiential rock bottom may easily lead back to subjectivism, as witnessed by operationism.

To sum up, we have proposed a contextual but not holistic theory of sense. The contextual nature of sense explains why one and the same symbol can be interpreted differently in different contexts and never fully outside some well delineated conceptual framework. But the matter of interpretation deserves another chapter. As a matter of fact we shall devote it the initial chapter of Part II of this work.

CONTINUED IN VOLUME 2: INTERPRETATION AND TRUTH

BIBLIOGRAPHY

Ajdukiewicz, K. (1935). Syntactic connexion. In S. McCall, Ed., *Polish Logic*. Oxford: Clarendon Press, 1967.

Ajdukiewicz, K. (1967a). Intensional expressions. *Studia Logica* **20**: 63–86.

Anderson, A. R. (1972). An intensional interpretation of truth-values. *Mind* **80**: 348–371.

Angel, R. B. (1970). Relativity and Covariance. Ph. D. thesis, McGill University.

Angel, R. B. (1973). Relativity and Covariance. In Bunge, Ed. 1973b.

Angelelli, I., Ed. (1967). *Gottlob Frege, Kleine Schriften*. Hildesheim: Georg Olms.

Arbib, M. A., Ed. (1968). *Algebraic Theory of Machines, Languages, and Semi-groups*. New York and London: Academic Press.

Aristotle. *De interpretatione*. In *Works*, X, ed. W. D. Ross. Oxford: Clarendon Press 1921.

Aristotle. *Analytica poteriora. Ibid.*

Arnauld, A. and P. Nicole (1662). *La logique, ou l'art de penser*. Paris: Flammarion 1970.

Ayer, J. A. (1959). *Logical Positivism*. Glencoe, Ill.: Free Press.

Bar-Hillel, Y. (1970). *Aspects of Language*. Jerusalem: Magnes Press.

Bar-Hillel, Y. and R. Carnap (1953). Semantic information. *Brit. J. Phil. Sci.* **4**: 147–157.

Bar-Hillel, Y., E. I. J. Poznanski, M. O. Rabin, and A. Robinson, Eds. (1961). *Essays on the Foundations of Mathematics*. Jerusalem: Magnes Press.

Bell, J. L. and A. B. Slomson (1969). *Models and Ultraproducts*. Amsterdam: North-Holland.

Blumenthal, L. M. and K. Menger (1970). *Studies in Geometry*. San Francisco: W. H. Freeman & Co.

Bolzano, B. (1837). *Wissenschaftslehre*, 4 vols. Sulzbach: Seidelsche Buchandlung. Abridged Engl. transl. by R. George, *Theory of Science*. Berkeley and Los Angeles: University of California Press 1972.

Bolzano, B. (1851). *Paradoxes of the Infinite*. Transl. D. A. Steele. London: Routledge & Kegan Paul, 1950.

Bourbaki, N. (1970). *Théorie des ensembles*. Paris: Hermann.

Braithwaite, R. B. (1962). Models in the empirical sciences. In E. Nagel *et al.*, Eds. (1962), pp. 224–231.

Brillouin, L. (1962). *Science and Information Theory*. New York: Academic Press.

Bunge, M. (1956). Survey of the interpretations of quantum mechanics. *Amer. J. Phys.* **24**: 272–286. Reprinted in Bunge (1959b).

Bunge, M. (1959b). *Metascientific Queries*. Springfield, Ill.: Charles C. Thomas Publ.

Bunge, M. (1961a). Laws of physical laws. *Amer. J. Phys.* **29**: 518–529. Repr. in Bunge (1963a).

Bunge, M. (1961b). The weight of simplicity. *Phil. Sci.* **28**: 120–149.

Bunge, M. (1962). *Intuition and Science*. Englewood Cliffs, N. J.: Prentice-Hall.

Bunge, M. (1966). Mach's critique of Newtonian mechanics. *Amer. J. Phys.* **34**: 585–596.

Bunge, M. (1967a). *Scientific Research*, 2 vols. New York: Springer-Verlag.

Bunge, M. (1967b). *Foundations of Physics*. New York: Springer-Verlag.

Bunge, M. (1967c). Analogy in quantum mechanics: from insight to nonsense. *Brit. J. Phil.*

Sci. **18**: 265–286.

Bunge, M. (1967d). Physical axiomatics. *Rev. Mod. Phys.* **39**: 463–474.

Bunge, M. (1967e). A ghost-free axiomatization of quantum mechanics. In M. Bunge, Ed., *Quantum Theory and Reality*, pp. 98–117. New York: Springer-Verlag.

Bunge, M. (1967f). The structure and content of a physical theory. In M. Bunge, Ed., *Delaware Seminar in the Foundations of Physics*, pp. 15–27. New York: Springer-Verlag.

Bunge, M. (1968). Problems and games in the current philosophy of natural science. *Proc. XIVth Intern. Congress of Philosophy* I: 566–574. Wien: Herder.

Bunge, M. (1969). What are physical theories about? *Amer. Philos. Quart. Monograph* **3**: 61–99.

Bunge, M. (1970a). Theory meets experience. In H. Kiefer and M. Munitz, Eds., *Contemporary Philosophic Thought*, **2**: 138–165. Albany, N.Y.: State University of New York Press.

Bunge, M. (1970b). Problems concerning inter-theory relations. In P. Weingartner and G. Zecha, Eds., *Induction, Physics, and Ethics*, pp. 287–325. Dordrecht: D. Reidel Publ. Co.

Bunge, M. (1971a). A mathematical theory of the dimensions and units of physical quantities. In M. Bunge, Ed., *Problems in the Foundations of Physics*, pp. 1–16. New York: Springer-Verlag.

Bunge, M. (1971b). Is scientific metaphysics possible? *J. Phil.* **68**: 507–520.

Bunge, M. (1972a). A program for the semantics of science. *J. Phil. Logic* **1**: 317–328. Reprinted in Bunge, Ed. (1973a).

Bunge, M. (1972b). Metatheory. In *Scientific Thought*. Paris–The Hague: Mouton/Unesco.

Bunge, M. (1973a). *Method, Model and Matter*. Dordrecht: D. Reidel Publ. Co.

Bunge, M. (1973b). *Philosophy of Physics*. Dordrecht: D. Reidel Publ. Co.

Bunge, M. (1973c). On confusing "measure" with "measurement" in the methodology of behavioral science. In Bunge, Ed. (1973b).

Bunge, M. (1973d). Meaning in science. *Proc. XVth World Congress of Philosophy* **2**: 281–286. Sofia.

Bunge, M., Ed. (1973a). *Exact Philosophy: Problems, Goals, and Methods*. Dordrecht: D. Reidel Publ. Co.

Bunge, M., Ed. (1973b). *The Methodological Unity of Science*. Dordrecht: Reidel.

Bunge, M. (1974a). The relations of logic and semantics to ontology. *J. Phil. Logic* **3**: 195–210.

Buridan, J. *Sophisms on Meaning and Truth [Sophismata]*. Transl. T. K. Scott. New York: Appleton-Century-Crofts 1966.

Campbell (1920). *Physics: The Elements*. Cambridge: Cambridge University Press. Repr.: *Foundations of Science*. New York: Dover Publications, Inc. 1957.

Carnap, R. (1936). Testability and meaning. *Phil. Sci.* **3**: 419–471; **4**: 1–40.

Carnap, R. (1939). *Foundations of Logic and Mathematics*. Chicago: University of Chicago Press.

Carnap, R. (1942). *Introduction to Semantics*. Cambridge, Mass.: Harvard University Press.

Carnap, R. (1947). *Meaning and Necessity*. Chicago: University of Chicago Press. Enlarged ed. 1956.

Carnap, R. (1952). Meaning postulates. *Phil. Studies* **3**: 65–73.

Carnap, R. (1956). The methodological character of theoretical concepts. In H. Feigl and M. Scriven, Eds., *Minnesota Studies in the Philosophy of Science*, I: 38–76. Minneapolis: University of Minnesota Press.

Carnap, R. (1958). *Introduction to Symbolic Logic and its Applications*. New York: Dover

Publications, Inc.
Carnap, R. (1961). On the use of Hilbert's ε-operator in scientific theories. In Bar-Hillel *et al.* Eds. (1961), 156–164.
Carnap, R. (1963a). Intellectual autobiography. In Schilpp Ed. (1963), pp. 3–84.
Carnap, R. (1963b). Replies and systematic expositions. In Schilpp Ed. (1963), pp. 859–1013.
Carnap, R. (1966). *Philosophical Foundations of Physics.* M. Gardner, Ed. New York: Basic Books.
Castonguay, C. (1971). Meaning and Existence in Mathematics. Ph. D. thesis, McGill University.
Castonguay, C. (1972). *Meaning and Existence in Mathematics.* Wien-New York: Springer-Verlag.
Chomsky, N. (1963). Formal properties of grammars. In Luce *et al.* 1963, Vol. II, pp. 323–418.
Chomsky, N. and G. A. Miller (1963). Introduction to the formal analysis of natural languages. In Luce *et al.* 1963, Vol. II, pp. 269–321.
Church, A. (1951). A formulation of the logic of sense and denotation. In P. Henle, H. M. Kallen and S. K. Langer, Eds., *Structure, Method and Meaning: Essays in Honor of Henry M. Sheffer*, pp. 3–24. New York: The liberal Arts Press.
Church, A. (1956). Propositions and sentences. In *The Problem of Universals: A Symposium*, pp. 3 11. Notre Dame, Ind.: University of Notre Dame Press.
Church, A. (1973/74). Outline of a revised formulation of the logic of sense and denotation. *Nous* **7**: 24–33, **8**: 135–156.
Chwistek, L. (1949). *The Limits of Science.* London: Routledge and Kegan Paul.
Cohen, L. J. (1962). *The Diversity of Meaning.* London: Methuen.
Cole, M. and I. Maltzman, Eds. (1969). *A Handbook of Contemporary Soviet Psychology.* New York and London: Basic Books, Inc.
Dummett, M. (1973). *Frege. Philosophy of Language.* London: Duckworth & Co. Ltd.
Feigl, H. (1950). Existential hypotheses. *Phil. Sci.* **17**: 35–62.
Feigl, H. (1958). The 'mental' and the 'physical'. In H. Feigl, M. Scriven, and G. Maxwell, Eds., *Minnesota Studies in the Philosophy of Science*, II: 370–497.
Feyerabend, P. K. (1970). Against method: Outline of an anarchistic theory of knowledge. In Radner and Winokur Eds. (1970) pp. 17–130.
Fraassen, B. C. van (1971). *Formal Semantics and Logic.* New York: Macmillan.
Frege, G. (1879). Begriffsschrift. Repr. in *Begriffsschrift und andere Aufsätze.* I. Angelelli, Ed. Hildesheim: Georg Olms 1964.
Frege, G. (1891). Funktion und Begriff. In Angelelli (1967).
Frege, G. (1892). Über Sinn und Bedeutung. In Angelelli (1967).
Frege, G. (1894). Rezension von: E. G. Husserl, Philosophie der Arithmetik. I. In Angelelli (1967).
Frege, G. (1895). Le nombre entier. In Angelelli (1967).
Frege, G. (1912). Anmerkungen zu P. E. B. Jourdain, The Development of the Theories of Mathematical Logic, etc. In Angelelli (1967).
Frege, G. (1969). *Nachgelassene Schriften.* H. Hermes, F. Kambartel and F. Kaulbach, Eds. Hamburg: F. Meiner.
Freudenthal, H. (1971). More about *Foundations of Physics. Foundations of Physics* **1**: 315–323.
Geach, P. T. (1962). *Reference and Generality.* Ithaca, N.Y.: Cornell University Press.
Geach, P. T. and M. Black, Eds. (1952). *Translations from the Philosophical Writings of Gottlob Frege.* Oxford: Basil Blackwell.

Ginsburg, S. (1966). *The Mathematical Theory of Context-Free Languages*. New York: McGraw-Hill.

Goodman, N. (1951). *The Structure of Appearance*. Cambridge, Mass.: Harvard University Press.

Goodman, N. (1961). About. *Mind N.S.* **70**: 1–24.

Goodman, N. (1968). *The Languages of Art*. Indiana and New York: Bobbs-Merrill Co.

Harris, Z. (1968). *Mathematical Structure of Languages*. New York: Interscience.

Hartnett, W. E. (1963). *Principles of Modern Mathematics*, 1. Chicago: Scott, Foresman, and Co.

Hartnett, W. E. (1970). *Principles of Modern Mathematics*, 2. Chicago: Scott, Foresman, and Co.

Heisenberg, W. (1955). The development of the interpretation of the quantum theory. In W. Pauli, L. Rosenfeld, and V. Weisskopf, Eds., *Niels Bohr and the Development of Physics*, pp. 12–29. London: Pergamon.

Heisenberg, W. (1958). The representation of nature in contemporary physics. *Daedalus* **87**: 95–108.

Hempel, C. G. (1945). Studies in the logic of confirmation. *Mind N.S.* **54**: 1–26, 97–121.

Hempel, C. G. (1950). A note on semantic realism. *Phil. Sci.* **17**: 169–173.

Hempel, C. G. (1958). The theoretician's dilemma. In H. Feigl, M. Scriven, and G. Maxwell, Eds., *Minnesota Studies in the Philosophy of Science* II: 37–98. Minneapolis, Min.: University of Minnesota Press.

Hempel, C. G. (1965). *Aspects of Scientific Explanation and Other Essays in the Philosophy of Science*. New York: The Free Press.

Henkin, L. (1953). Some notes on nominalism. *J. Symbol. Logic* **18**: 19–29.

Henkin, L., P. Suppes and A. Tarski, Eds. (1959). *The Axiomatic Method*. Amsterdam: North-Holland Publ. Co.

Hilbert, D. (1900?). Letter to G. Frege. In Angelelli (1967).

Hilbert, D. and W. Ackermann (1950). *Principles of Mathematical Logic*. New York: Chelsea Publ. Co.

Hilbert, D. and P. Bernays (1968). *Grundlagen der Mathematik* I, 2nd ed. Berlin-Heidelberg-New York: Springer-Verlag.

Hill, T. E. (1971). *The Concept of Meaning*. New York: Humanities Press.

Hintikka, J. (1968). Varieties of information. In B. v. Rootselaar and J. F. Staal, Eds., pp. 311–332.

Hintikka, J. (1970). On semantic information. In W. Yourgrau and A. D. Breck, Eds., *Physics, Logic, and History*, pp. 147–172. New York and London: Plenum Press.

Katz, J. J. and J. A. Fodor (1963). The structure of a semantic theory. *Language* **39**: 170–210.

Kneale, W. and M. (1962). *The Development of Logic*. Oxford: Clarendon Press.

Kneale, W. (1972). Numbers and numerals. *Brit. J. Phil. Sci.* **23**: 191–206.

Kraft, V. (1970). *Mathematik, Logik and Erfahrung*, 2nd ed. Wien and New York: Springer-Verlag.

Kuhn, T. (1962). *The Structure of Scientific Revolutions*. Chicago: University of Chicago Press.

Lambert, K. and B. C. v. Fraassen (1970). Meaning relations, possible objects, and possible worlds. In K. Lambert, Ed., *Philosophical Problems in Logic*, pp. 1–19. Dordrecht: D. Reidel Publ. Co.

Lawvere, F. W. (1966). The category of categories as a foundation of mathematics. *Proceedings of the Conference on Categorical Algebra* (La Jolla 1965), pp. 1–20. Berlin-Heidelberg-New York: Springer-Verlag.

Leonard, H. S. (1967). *Principles of Reasoning*. New York: Dover Publications Inc.
Lewis, C. I. (1944). The modes of meaning. *Philosophy and Phenomenological Research* **4**: 236–249.
Lewis, C. I. (1951). Notes on the logic of intensions. In P. Henle *et al.*, Eds., *Structure, Method and Meaning*. New York: Liberal Arts Press.
Linsky, L. (1967). *Referring*. London: Routledge and Kegan Paul.
Linsky, L., Ed. (1952). *Semantics and the Philosophy of Language*. Urbana: University of Illinois Press.
Luria, A. R. (1969). Speech development and the formation of mental processes. In Cole and Maltzmann, Eds. 1969, pp. 121–162.
Luce, R. D., R. R. Bush, and E. Galanter, Eds. (1963). *Handbook of Mathematical Psychology*, Vol. II. New York: John Wiley and Sons, Inc.
MacKay, D. M. (1969). *Information, Mechanism and Meaning*. Cambridge, Mass.: M.I.T. Press.
Marcus, S. (1967). *Algebraic Linguistics: Analytical Models*. New York: Academic Press.
Marhenke, P. (1950). The criterion of significance. In Linsky, Ed. (1952), pp. 139–159.
Montague, R. (1970). Pragmatics and intensional logic. *Synthese* **22**: 68–94.
Morris, C. (1938). *Foundations of the Theory of Signs*. Chicago: University of Chicago Press.
Nagel, E. (1950). Science and semantic realism. *Phil. Sci.* **17**: 174–181.
Nagel, E., P. Suppes, and A. Tarski, Eds. (1962). *Logic, Methodology and Philosophy of Science*. Stanford: Stanford University Press.
Neumann, J. v. (1932). *Mathematische Grundlagen der Quantenmechanik*. Berlin: J. Springer.
Osgood, C. E., G. J. Suci, and P. H. Tannenbaum (1957). *The Measurement of Meaning*. Urbana, Ill.: University of Illinois Press.
Patton, T. E. (1965). Some comments on "About". *J. Phil.* **62**: 311–325.
Popper, K. R. (1963a). The demarcation between science and metaphysics. In Schilpp, Ed. (1963), pp. 183–226.
Popper, K. R. (1963b). *Conjectures and Refutations*. New York: Basic Books.
Popper, K. R. (1966). A theorem on truth-content. In P. K. Feyerabend and G. Maxwell, Eds., *Mind, Matter, and Method: Essays in Philosophy and Science in Honor of Herbert Feigl*, pp. 343–353. Minneapolis: University of Minnesota Press.
Przelecki, M. (1969). *The Logic of Empirical Theories*. London: Routledge and Kegan Paul.
Putnam, H. (1970). Is semantics possible? *Metaphilosophy* **1**: 187–201.
Putnam, H. (1971). *Philosophy of Logic*. New York: Harper & Row Publ.
Putnam, H. (1973). Meaning and reference. *Journal of Philosophy* **70**: 699–711.
Quine, W. V. (1952). *From a Logical Point of View*. Cambridge, Mass.: Harvard University Press.
Quine, W. V. (1966). *The Ways of Paradox and Other Essays*. New York: Random House.
Quine, W. V. (1971). Epistemology naturalized. *Proc. XIVth Intern. Congress Phil.* VI: 87–103. Wien: Herder.
Radner, M. and S. Winokur, Eds. (1970). *Minnesota Studies in the Philosophy of Science* IV. Minneapolis: University of Minnesota Press.
Robinson, A. (1963). *Introduction to Model Theory and to the Metamathematics of Algebra*. North-Holland Publ. Co.
Robinson, A. (1966). *Non-standard Analysis*. North-Holland Publ. Co.
Rootselaar, B. van, and J. F. Staal, Eds. (1968). *Logic, Methodology and Philosophy of Science* III. Amsterdam: North-Holland Publ. Co.

Rosen, R. (1958). The representation of biological systems from the standpoint of the theory of categories. *Bull. Math. Biophys.* **20**: 317–341.

Rozeboom, W. (1962). The factual content of theoretical concepts. In H. Feigl and G.

Rosenbloom, P. (1950). *The Elements of Mathematical Logic.* New York: Dover.

Maxwell, Eds., *Minnesota Studies in the Philosophy of Science* III: 273–357. Minneapolis, Min.: University of Minnesota Press.

Rozeboom, W. (1970). The crisis in philosophical semantics. In Radner and Winokur, Eds. pp. 196–219.

Russell, B. (1918). The philosophy of logical atomism. *The Monist* **28**: 495–527; **29**: 32–63, 190–222, 345–380.

Russell, B. (1940). *An Inquiry Into Meaning and Truth.* London: George Allen & Unwin.

Russell, B. (1948). *Human Knowledge: Its Scope and Limits.* London: George Allen & Unwin.

Russell, B. (1959). *My Philosophical Development.* London: George Allen & Unwin.

Ryle, G. (1949). *The Concept of Mind.* London: Hutchinson.

Schaff, A. (1962). *Introduction to Semantics.* Oxford: Pergamon Press.

Schilpp, P. A., Ed. (1963). The Philosophy of Rudolf Carnap. La Salle, Ill.: Open Court; London: Cambridge University Press.

Scott, D. (1970). Advice on modal logic. In K. Lambert, Ed., *Philosophical Problems in Logic.* Dordrecht: D. Reidel Publ. Co.

Searle, J. R. (1969). *Speech Acts: An Essay in the Philosophy of Language.* Cambridge: University Press.

Seshu, S. and M. B. Reed (1961). *Linear Graphs and Electrical Networks.* Reading, Mass.: Addison-Wesley.

Shannon, C. and W. Weaver (1949). *The Mathematical Theory of Communications.* Urbana, Ill.: University of Illinois Press.

Sharvy, R. (1972). Three types of referential opacity. *Phil. Sci.* **39**: 152–161.

Shwayder, D. S. (1961). *Modes of Referring and the Problem of Universals.* Berkeley and Los Angeles: University of California Press.

Suppe, F. (1971). On partial interpretation. *J. Phil.* **68**: 57–76.

Suppes, P. (1960). *Axiomatic Set Theory.* Princeton, N.J.: D. Van Nostrand.

Suppes, P. (1967). *Set-theoretical Structures in Science.* Stanford University: Institute for Mathematical Studies in the Social Sciences.

Suppes, P. (1969). *Studies in the Methodology and Foundations of Science.* Dordrecht: D. Reidel.

Suszko, R. (1967). An essay in the formal theory of extension and intension. *Studia Logica* **20**: 7–28.

Tarski, A. (1956). *Logic, Semantics, Metamathematics.* Oxford: Clarendon Press.

Tarski, A., A. Mostowski, and R. M. Robinson (1953). *Undecidable Theories.* Amsterdam: North-Holland Publ. Co.

Vaihinger, H. (1920). *Die Philosophie des Als Ob*, 4th ed. Leipzig: F. Meiner.

Waddington, C. H. (1967). Discussion of Eden's paper. In P. S. Moorhead and M. M. Kaplan, Eds., *Mathematical Challenges to the Neo-Darwinian Interpretation of Evolution.* Philadelphia: The Wistar Institute.

Watson, W. H. (1950). *On Understanding Physics.* New York: Harper & Brothers.

Weingartner, P. (1972). Die Fraglichkeit der Extensionalitätsthese und die Probleme einer intensionalen Logik. In R. Haller, Ed., *Jenseits von Sein und Nichtsein.* Graz: Akademische Druck-u. Verlagsanstalt.

Weingartner, P. (1973). A predicate calculus for intensional logic. *J. Phil. Logic* **2**: 58–141.

Whitehead, A. N. and B. Russell (1927). *Principia Mathematica* I. 2nd. ed. Cambridge: Cambridge University Press.

Wigner, E. P. (1970). Physics and the explanation of life. *Foundations of Phys.* **1**: 35–45.

Williams, D. C. (1937). The realistic interpretation of scientific sentences. *Erkenntnis* **7**: 169–178, 375–382.

Williams, D. C. (1966). *Principles of Empirical Realism.* Springfield, Ill.: Charles C. Thomas.

Williams, G. C. (1966). *Adaptation and Natural Selection.* Princeton, N.J.: Princeton University Press.

Williams, M. B. (1970). Deducing the consequences of evolution: A mathematical model. *J. Theoret. Biol.* **29**: 343–385.

Wojcicki, R. (1966). Semantical criteria of empirical meaningfulness. *Studia Logica* **19**: 75–102.

Yourgrau, W. and A. D. Breck, Eds. (1970). *Physics, Logic, and History.* New York and London: Plenum Press.

Zinov'ev, A. A. (1973). *Foundations of the Logical Theory of Scientific Knowledge (Complex Logic).* Dordrecht: D. Reidel.

INDEX OF NAMES

INDEX OF SUBJECTS

The companion of this book is Volume 2 of the

Treatise on Basic Philosophy

INTERPRETATION AND TRUTH

TABLE OF CONTENTS